Trump's Wall

The Wall That Built Itself
And Made Money Doing It
A Transcontinental Corridor for a Superconducting Power
Line
Creating Jobs and Stimulating the Free Market Economy
Trump's American Version of the Great Wall of China
Visionary, Enduring

The Wall That Made America Great Again

http://www.thewallthatbuiltitself.com

Published by:
SAN 299-2701
Glenhaven Publications
Seattle, Washington

ISBN   978-0-9984627-9-0 (paperback)

Dedication

To my Mother
for her encouragement, patience and council

Table of Contents

Preface

The United States has a problem with people entering illegally.

Perhaps the first thing to do is ask "Why?"

People uproot themselves, suffer hardships and brave dangers to come to the United States. Why? Because each believes it will bring them the opportunity for a better life.

This driving force is held at the individual level, it is very personal.

Each individual acts upon their personally held belief that there is a disparity between the opportunities they have, compared to the opportunities they believe they could have. They assume the risks with the hope of reward.

Conversely we may ask, "Why do very few people leave the United States for other countries?" Because most people in the United States believe they have better opportunities where they are.

This is a deeply held personal belief, not a supposition, and people act accordingly.

As long as people hold this belief they will not stop coming, no matter what they are told. Telling people to not believe what they can see is like demanding someone with whooping cough to stop coughing.

It seems there are two ways to deal with this problem.

- One way is to make illegal entry very difficult.

- Another way is to remove the very reason for illegal entry - by creating opportunities for people outside of the United States.

To address the first point, a wall has been proposed - and walls work. However, a simple wall is a static, lifeless thing. It must be supported by external forces. It creates no value, it has few vested interests. It is an expense in the financial ledger. Over time it will have many more forces trying to tear it down and subvert it, than maintain and improve it.

Also, a wall is seen by some as an expense in the emotional ledger. A representation that we, as a nation, are selfish and fearful. Thus it could be seen by some as embarrassing and damaging to our national self-image.

But there is no question that all sovereign nations have the right and the obligation to control their borders, just as we each have the right and obligation to control our own homes.

As to the second point, can we design in such a way that both the United States and Mexico benefit? Can we design a solution that exemplifies the problem solving ingenuity our country was built on?

Is there a way to create a border environment that protects our sovereignty, produces wealth for us and provides economic opportunities in Mexico? Is there a way to create a wealth-producing environment where, instead of us demanding Mexico pay for a wall, they would ask to invest?

I think there is.

All the public discussion I am aware of is based on the unspoken assumption of a wall, purpose built to control immigration.

Let us change the basic assumption of a wall. Try thinking about the wall as a much greater project that serves a much greater need for humanity.

To explore and utilize a new environment. To project ourselves beyond our world, beyond our world view, beyond how we now view ourselves. ***To become a space fairing race.*** Not to visit, not to live there just yet, but to have economically useful access to space.

This book is intended to discuss how the staged economic development of the wall could lead toward those greater goals.

The Grand Vision

Upon first hearing of the idea of a southern border wall, I did not put much thought into it. But, over time, I have come to believe that it has the potential to be the means to uplift the human condition, to change mankind's self-image.

As an engineer I became quite excited about the prospect of a grand project of our time. After all, many development projects these days are computer code, and whereas many are very useful and of great power they are ethereal, not of the physical world. It has been generations since we attempted grand things. The Erie Canal 1825, Great Northern Railway 1889, The Panama Canal 1914, Hoover Dam 1936, Grand Coulee Dam 1942, Saturn V Rocket 1967, Apollo Lunar Mission 1969. It has been a while since we have attempted a physical project with the character to capture our society's imagination.

So, drifting off to sleep one night about a year ago, I found myself mulling over how to go about building a 2,000 mile long wall. Commentators had pointed out that in various sections of our southern border the terrain is quite challenging. This lead me to recall railway construction methods of the past. With the need to push rail through miles of undeveloped hostile terrain they used trains on the newly laid rail to bring material to the railhead to lay more rail ahead. This obviated the time and expense of building a parallel road infrastructure.

Ah, use the wall to transport material to build the wall.

As is my want, one idea after another tumbled out of my imagination until a grand vision stood before me.

I even outlined my belief in the possibilities in a letter to President Trump.

Letter to President Trump

Mark Olson

Seattle, WA

January 21, 2017

The Wall That Built Itself
And Made Money Doing It
A Transcontinental Corridor for a Superconducting Power Line
Creating Jobs and Stimulating the Free Market Economy
Trump's American version of the Great Wall of China
Visionary, Enduring

The President
The White House
1600 Pennsylvania Avenue NW
Washington, DC 20500

Dear Mr. President:

You have told us you want to build a wall to protect our nation's sovereignty. Who will pay for it? Your construction development history shows an emphasis on cost controls and a focus on generating income early in a project. Why not build a wall the same way? Design it to pay for itself. Design it to create jobs and spur private enterprise. Design it to address our nation's energy crisis. Design it in such a way that Mexico wants it and benefits from it. Design a wall that exemplifies the problem solving ingenuity our country was built on.

A wall is a 20 or 30 foot tall structure that divides space. Two parallel walls ... can be a building. Imagine a 2,000 mile long building that provides the boarder security of a wall while at the same time serves as a new secure transnational corridor for our nation's overworked and outdated electrical grid.

Such a continent spanning building could provide a corridor to house a new transnational superconducting power line, hardened against electromagnetic pulse and secure from attack. In fact such a structure may be one of our last viable remaining transcontinental corridors, through which could run the backbone of an electrical power generating solar panel array.

Solar panel power generating arrays could be built along the new power line by adjacent land owners thus promoting free market participation and investment. As long as the government played a limited role, private companies would be attracted to the new economic opportunities and thousands of new jobs would be created.

Mister President, the wall you propose to build has created a unique opportunity in history to create an environment, an economic ecosystem, that encourages people to cooperate and be productive based on individual free choice and free will.

A project of this magnitude would require the awesome power of government and the boundless energy of a free people. We Americans have always dreamed big, it bubbles out of our optimism, we can't help ourselves.

This vision is flashy, bold, daring and terrifying. It is fraught with hidden danger and a high chance of failure, but oh what success could look like!

Most Respectfully:

Mark Olson
B.S. Mechanical Engineering

*[Original attachment to the letter including
"Further Thoughts" and "Possibilities"
follows on the next pages]*

Further Thoughts

- Too much government will make the project fail
  - Government as referee
  - Government owned but leased on the open market
    - The Old Post Office Building as an example?
- Create income quickly by smart construction sequencing
  - First build a minimum double wall with a conventional electrical power line
  - Create income by tying electrical power generating solar panel arrays into the power line
  - Then install a transnational superconducting power line
  - etc
- Security - can be efficiently provided for the building, everything it contains and its environment
  - Border
  - Power Grid
  - Power Generation
  - Transportation
  - Etc.
- Politics - By bringing prosperity you gain political strength
  - A building/power corridor is a prosperity generator and thus brings you political power
  - A simple wall is an expense and a political weapon to be used against you

- International relations
  - What is good for us does not have to be bad for them.
  - Mexicans don't have to pay for the wall (at least not entirely), it pays for itself

- Stimulate economic activity in Mexico
  - solar panel electrical generation etc.
- Unintended consequences - There is no such thing as as sure thing.
  - There are always unintended consequences, but left to the free economy and ingenuity of the American people these are predominantly constructive and positive.

Possibilities

- Desalination plant at each end
  - Provide water for building needs and adjoining land irrigation
- Magnetic Levitation (Maglev) Train
- Hyperloop
  - 3 hours to transit the continent (600-700 mph)
  - Freight - Attract FedEx, Amazon, etc distribution centers
  - People transportation - practicable to commute hundreds of miles to work
- Sea Port at each end
  - Properly designed it could compete with the Panama canal
  - Locks lift entire ships onto maglev carriages
    - Transcontinental transit of entire ships in less than one day
  - Compete with existing sea ports
    - Locks need be the only infrastructure at the ports

- Entire ships brought into the building to be off/on loaded
  - Direct access to distribution centers located along the building
  - Avoid the congestion of port city road systems
- The entire building can be thought of as a giant material handling system
  - Overhead cranes, conveyers, etc.
- Ship size no longer limited by existing Panama canal locks
- National security improved
- Manufacturing attracted by
  - Transportation
  - Power
  - Climate controlled spaces
  - Advanced material handling (i.e. overhead cranes, conveyers, etc.)
- Nuclear power plants
  - Modular
  - Manufactured and operated within the building
  - Tie into electrical grid
  - Secure
- A gateway to space
  - Spaceship manufacturing
  - Rail Gun Spaceship Launch facilities for modular rockets
  - Launch rate exponentially higher that at present
- Building sized modules movable internally in the building
  - Living space
  - Factories
- Who knows, the possibilities of American ingenuity are endless!

[End of Attachment to Letter]

A Reply

After several months Patrick G. Carrick, Ph.D., Director, Homeland Security Advanced Research Projects Agency replied to me by letter. In his letter he very graciously thanked me for my efforts and ideas.

Streamlining my original letter in order to keep it short enough for people to be willing to read had, I felt, not conveyed the full scope of the enormous potential of President Trump's Wall. Therefore, I sent Dr. Carrick a second letter with my ideas more fully explained.

Letter to Patrick Carrick Ph.D.

Mark Olson

Seattle, WA

May 02, 2017

The Wall That Built Itself
And Made Money Doing It
A Transcontinental Corridor for a Superconducting Power Line
Creating Jobs and Stimulating the Free Market Economy
Trump's American version of the Great Wall of China
Visionary, Enduring

Patrick G. Carrick, Ph.D.
Director, Homeland Security Advanced Research Projects Agency
U. S. Department of Homeland Security
Washington, DC 20528

Dear Dr. Carrick:

Thank you for your letter of April 24, 2017. At the risk of imposing on your kindness I though it might be useful to expand on some of the ideas I mentioned in the "Possibilities" section of my previous letter (January 21, 2017).

Respectfully:
Mark Olson
B.S. Mechanical Engineering

[Original attachment to the letter including
fifteen explanatory topics
follows on the next nine pages]

I think of the Trump Wall as a Building and the Building as a giant Material Handling Corridor, replete with overhead cranes, conveyers, trains, power distribution, etc.

I think of a building 40 stories tall, 40 stories deep, two to four miles wide and 2,000 miles long, energized with the raw power of robotized material handling and manufacturing and filled with the boundless spirit of American free-enterprise.

Build The Building In Stages
- Build in stages to generate income as soon as possible.
- I think it would be highly inadvisable for the government to try to build it all at once. Not a *fait accompli* and then look for people to use it.
- I think it is vital to build it in stages by private investment, companies and people with focus on creating income early and often. Getting as many interested parties involved as is practicable will make the project more robust. The more people making money the more people will want it.
- Once the first stage is completed, spanning the continent and making income, the building becomes both the transportation/material handling corridor and the validation to attract investment capital to build the next layer of the Building, etc.

Use The Building To Build The Building
- The terrain is difficult in places. Instead of building roads beside a wall to move supplies. Use the Building to transport material to build the Building. This is how railroads were built in centuries past. They used trains on the newly laid rail to bring material to the railhead to lay more rail ahead.

Super-Conducting Transcontinental Power Line
- Secure and hardened against attack. Perhaps our best, our only practicable corridor.
- The Building would provide protection and easy access for repair and upgrading. No need to bury or elevate it. The Building would provide environmental control, over-head cranes, component manufacturing, transportation of material and people, etc.

Solar Panel Power Generating Arrays
- Encourage and provide private investment opportunities in both USA and Mexico.
- Solar panel arrays on land bordering both sides of the Building, 20-30 feet above ground with water pipes under them to "rain" irrigation on the land below for crops and pasture land.
- For solar panels of present day technology and efficiency, a rough calculation shows (please double check) that an array 20 miles wide by 2,000 miles long would generate the equivalent of 100% of the present USA energy consumption.

Desalination Plants

- One at each end of the Building drawing water from the oceans.
- Feed the Building's and adjoining land's irrigation fresh water needs.
- Use energy generated in and around the building to power the desalination.
- Irrigating land on both sides of the Building will produce a multi-billion dollar agricultural zone. Thus providing both USA and Mexico with more investment opportunities.

HyperLoop

- People transportation/shipping container transportation (700 mph).
- Live/work in the Building. A 1,000 mile daily commute would be practicable.
- https://en.wikipedia.org/wiki/Hyperloop

Sea Port at Each End - MagLev Train Competition to the Panama Canal

- Use MagLev technology for transcontinental shipping.
- Have locks in each ocean to elevate entire container ships onto maglev carriages and move them through the Building to the other ocean (two days to transit continent at 50 mph).
- Ship cargo exchange done inside the building as needed. And/or

- Off-load shipping containers into Building to hyper-loop (3 hours to transit continent at 700 mph) on-load to ship in other ocean.
- Robotized shipping container on-load and off-load.
- Building being one large Material Handling System would direct incoming shipping containers to distribution centers built along the USA side of the Building (FedEx, Amazon, Maersk, Hyundai, etc.) and out-going containers to ships in either ocean.
- The Building would be competition to the Panama Canal, but so much more.
- Better national security - Every container can be scanned at point of entry.
- Better national security - Protects our nation's sovereignty by not relying on other nations to providing shipping access to both USA coasts.
- Moves port facilities away from limited, weather exposed, expensive coastal locations.
- The size of the locks of the Panama canal are limiting the size of ships. With this system, accommodating larger ships in the future would require relatively minor infrastructure changes.

Shipping - International Distribution Centers

- Large transit centers built along the USA side of the Building (FedEx, Amazon, Maersk, Hyundai, etc.) built to handle shipping containers and tied into the Building's HyperLoop and MagLev Material Handling Corridor and existing distribution systems of the USA (truck, train, airplane, etc.).

Factory - Residential

- If the Building is thought of as framework that has a transportation/material handling corridor and, on the USA side, a wall that is a bank of standard sized receptacles (like a bank of post office boxes).
- Standard sized Boxes fit into these receptacles/cubby holes.
- To move a Box, it is drawn out onto the transportation corridor, moved down the Building and inserted into a vacant receptacle.
- The standard Box size could be, say, a cube 200 feet on a side, basically a 20 story building. These Boxes could be built out as needed; residential, office, entertainment, manufacturing, industrial, etc.
- Some of the standard size Boxes could themselves be frameworks of receptacles that receive a number of smaller standard sized Boxes that could be built out to the customer's/owner's needs. Thus a person or small business could own their own home or office and move it at will anywhere along the wall.
- Standard quick-connect connections for water, drain, communications and electrical power would make moving inexpensive and quick.
- This means that any home or business could easily move elsewhere in the Building.

Nuclear Power

- http://www.usatoday.com/story/money/columnist/2017/01/22/compact-prefab-power-plants-may-revive-nuclear-option/96839164/
- NuScale Power, Oregon
- Modular self-contained nuclear cells ~50 megawatts
- The manufacturing facility for building these modular self-contained nuclear cells could be in the building.
- This will provide for easy, safe and secure transportation of the modular nuclear cells.
- The modular nuclear cells could be installed in a secure basement throughout the Building's length and tied into the super-conducting power line.
- Repair/refurbish/upgrades to the modular nuclear cells could be done in the Building by moving the factory to a cell or by moving a cell to the factory.
- Modular nuclear cells could be constantly updated to the latest technology by companies that own and operate them.
- The Building leases them space and provides security.

Gateway To Space - Rail Gun - Space Launch

- **StarTram** is a proposal for a maglev space launch system.
- *https://en.wikipedia.org/wiki/StarTram*
- MagLev rail gun(s) built along the top of the building for launching rockets.
- Power for launches generated in and around the Building.
- Design, build modular, container like, rockets in the Building.

- Manufacture and/or transport payloads to rockets in the Building.
- Much lower cost per pound of payload to orbit then present capabilities.
- Much faster launching cycle then present capabilities.

Security

- MagLev mounted weapons platforms that can move along the length of the wall.
- Some sections/levels would be high security; modular nuclear cells, communications, power distribution, etc.
- I think it would be counter-productive to have the entire building a federal government high security facility.

Governance

- I think each state and county should have jurisdiction over the part of the Building in its territory, with as much private ownership as possible.
- As mentioned above in the "Factory - Residential" section any home or business could be easily moved elsewhere in the Building. This makes it difficult for people and companies to be held hostage by any one jurisdiction because they can just move, lock, stock and barrel, if they are unhappy. This forces local governments to compete for residents and businesses. Thus limiting the power of government.
- There must be federal oversight of some aspects of the Building. Immigration of course. Also, the nation can not be held hostage to any small jurisdiction that wants to, say, interrupt transcontinental transportation, power,

water, communications, etc. through their section of the Building.

Build in Mexico - Leverage

- If there is too much resistance to build the Building in the USA, threaten to build it in Mexico.
- Get Mexico to set aside a 100 mile wide swath at the border.
- Make this a free economic zone, like Hong Kong.
- It will still belong to Mexico, pay Mexico taxes, but the zone will have its own government and free market laws.

Conclusion

Please notice, nowhere in this submission is controlling our border mentioned. That is a side benefit of this economic engine.

Over the past decades there have been many good ideas that have all lacked one critical thing...where to put them. Trump's Building fills this need for all of them. More than that, located together they mutually support each other so the whole is greater than the sum of its parts.

- Transcontinental Super-Conducting Power Line - We need this.
- Solar Panel Power Generation - With a power line to get the power where it is needed.
- HyperLoop - Not just a fun thing, the Building really needs it to be effective.
- MagLev Trains - A Panama canal alternative.

- Desalination Plants - Reliable/renewable fresh water supply for our Southwest.
- Nuclear Power - Safe, secure, scalable, renewable.
- Space Launch System - Economically able to move enough mass into orbit to matter.

At present the public discourse is "How much will Trump's wall cost?" Build his "wall" as an economic engine and the world will be pounding at our door demanding "How much can I invest?"

Incidentally, job creation in Mexico, due to solar panel power generation and fresh water land irrigation, would encourage Mexicans to stay in, and return to, Mexico.

[End of Attachment to Letter]

True Potential

But do these letters adequately express this opportunity's true potential? These are merely cool things we could do but the letters do not capture the reasons for doing them.

All public discussion I am aware of is based on the unspoken assumption of a wall, purpose built to control immigration. To say "add solar panels" is just dressing up a wall, the basic assumption being that it is merely a wall.

Let us change the basic assumption of a wall to thinking about this as a much greater project that serves a much greater need for our society.

To explore and utilize a new environment. To project ourselves beyond our world, beyond our world view, beyond how we now view ourselves. **To become a spacefaring race.** Not to visit, not to live there just yet, but to have economically useful access to space.

But, Why go to space? Not to space in general nor travel to other planets, but specifically to Earth orbit. What can we possibly gain from that? We don't go as tourists, we go to work, to manufacture, to produce. For things can be made in space that cannot be made on Earth.

In the nearly zero gravity of space the physical forces that dominate are different.

- Diffusion dominates the mixing of dissimilar materials and gases; no sedimentation, controlled convection. Thus large and nearly perfect crystals are possible, for jet turbine blades, etc.

- The vacuum of space is ultra-clean. Allowing for very pure materials. Vapor deposition of extreme purity is possible, for example: for super hard coatings on machine tool cutters, 3D printing at micron level accuracy, microelectronics, etc.

- Surface tension dominates in microgravity. This allows for perfect spheres to be made, for example: for ball bearings, etc.

- Temperature extremes of space are physically very close together. Focused sunlight can produce molten steel just fractions of an inch away from the nearly absolute zero of ink-black shade. This allows for production of super-strong amorphous glass-like materials, thermoelectric power generation, and driving the flow of liquids and gasses.

In light of the fascinating possibilities I have come to believe that President Trump's Wall is a unique opportunity in history to transform the human condition. To transform the very way humanity sees itself.

When was the last time we, as a people, were galvanized into action, captivated by a grand vision? Pushed beyond ourselves for some grand design. Drawn to something greater than ourselves purely for the pleasure of it. For the opportunities, surely, but just for the doing of it. When was the last time we were drawn outside ourselves, held spellbound by the very real human drama of daring and risk and tragedy and triumph.

The Panama Canal captivated the world's attention. People traveled for days to attend the Eire Canal grand opening. Precisely at midnight on a Sunday morning Grand Central Station opened its doors for the first time, within 16 hours more than 150,000 people had visited. My Grandparents traveled 250 miles to attend the opening of the Grand Coulee Dam. The world was galvanized by the space race. People's emotions were rent by the human struggle and suffering and risk and triumph of the lunar landings. People cared.

In the past half century various U.S. presidents have tried to capture people's imagination to do something tangible but it has been contrived and as such lacked the necessary authenticity. No one had a reason to care.

But this is real. We are ready. We have the technology. Most importantly, we have a reason to go to space. This is the next giant leap for mankind.

Yes, there will be success and tragedy, triumphs and setbacks for, to engineer is human. We are ready, technologically. The question is, do we have the will to dare great things and the determination to see them through?

The intent of this book is to elaborate and explain this vision of what could be.

Trump's Wall has the potential to advance mankind because the sum would be greater than its parts. Each part alone would be impressive but acting as a whole, human advancement would become inevitable.

This is what this book is about.

In The Beginning

In a project of this scope and magnitude it is vital to keep foremost in every decision that we are building something that creates prosperity and provides opportunities. A project, at every stage, that produces more than it consumes, that costs less than it makes. To do this, it would be imprudent to try to finance and build it all at once. Build in stages. Where each stage justifies itself and lays the groundwork for those to come after. This is the sure way of getting many vested interests fighting for its completion and survival.

With that in mind, let us begin.

Transcontinental Superconducting Power Line

That our national power grid is out of date and vulnerable to attack has been much discussed over the past several decades. But most of this discussion goes on in engineering and technical publications and forums out of the main stream of society. However, from time to time we, the average citizen, have been rudely awakened to our vulnerability by geographically large power interruptions, brown-outs, etc.

This comes as no surprise when one considers that much of our nation's power transmission and distribution lines were installed in the 1950's and 60's. Power lines that were designed to have 50 year life spans. We also have over half a million miles of high-voltage transmission lines that are pushed routinely to full capacity. From this description it seems catastrophic failure is more likely from within rather than from foreign attack.

Why install power transmission lines in Trump's Wall? What advantages would that bring?

- Protection from the elements
- Easy access for maintenance and improvements
- Security
- They would force the wall to be continuous.

Various proposals have been made over the years to improve our present grid but it is fiendishly difficult and expensive to upgrade existing systems.

One of the major roadblocks has been securing right-of-ways, the physical land necessary for ambitious projects. Trump's Wall may be one of our last viable remaining transcontinental corridors.

A point in favor of installing power lines within the wall is that power lines must be continuous, obviously. Thus transcontinental power lines would be a built-in reason for the wall to be continuous, without breaks.

One of the interesting things about electricity is "transmission losses". Transmission losses are electrical losses due to simply running electricity through wire. An example is an element on a common electrical stovetop. The stovetop element is designed in such a way to intentionally create great electrical transmission losses which are manifest as heat, seen as a red-hot glowing element. And yes, running too much current through the wires of your house will cause them to become red-hot and burn your house down, which is why your house has circuit breakers and/or fuses to prevent this.
But I digress.

Transmission losses increase with transmission distance. They can be reduced by greatly increasing the transmission voltage but losses are a consequence of physical law. In a national electrical grid with distances of many miles, transmission losses can amount to 10% in some cases.

Something on the order of 6% of all electricity generated in the US is lost to transmission losses. Not too bad, until we consider that represents a monetary loss in excess of $20 billion per year. Consider the distance from your home to the nearest power plant. As most of us would prefer to not live next door to a power plant we are willing to pay the transmission losses.

All is not lost. Heike Onnes, a Dutch physicist, discovered that some metals lose all electrical resistance at temperatures close to absolute zero, aptly called superconductivity. Which is to say that electricity suffers virtually zero transmission losses over infinite distance. This was in 1911. Since then advances have been made. In 2001, some smart Japanese guys discovered that, at the positively balmy temperature of 39 degrees above absolute zero (39 K), magnesium diboride (MgB_2) becomes superconductive. This inexpensive compound, and high temperature, made superconducting economically practicable, possibly. The city of Essen, Germany installed a short, one kilometer, superconducting power line in 2014. To date they seem happy with it.

Graphene is a one atom thick sheet of carbon atoms with a theoretical superconductive capability. To date no one has figured it out but there have been some recent discoveries.
> https://www.sciencealert.com/graphene-s-superconductive-power-has-finally-been-unlocked-and-it-s-crazier-than-we-expected

This is not to expect graphene power transmission lines tomorrow, who knows it may never pan-out. The point is, technology is advancing and it will be of great advantage to have the built-in infrastructure of the wall (cranes, transportation, lighting, etc.) to upgrade the transcontinental transmission lines as technology improves.

It is not my intention to insist on "superconducting" transcontinental power transmission lines. I do not know enough to calculate the best solution. What we need are transcontinental power transmission lines.

Why? First of all, it is a given that we need to replace our aging infrastructure. But we have another problem. The linking of power generating facilities to population centers, which tend to be at great distances from each other.

The American Southwest is ideal for solar and wind power generation. So why is it not covered with them? Because it is remote and sparsely populated. It is an ideal location to build these kinds of power generation facilities, but there is no way to economically get the power to those who will use it. There are no power lines and no right-of-ways to run them. Making Trump's wall a conduit for a transcontinental power transmission line would improve the viability of these power generation projects. It would tie these power generation sources into the national grid.

Our electrical power grid needs to balance itself. The supply needs to closely balance the demand. If there is much more electricity being produced than being used, safety devices on power plants shut them down to protect them from self-destructing. This can, in some cases, cause a cascade of shut-downs which results in total power loss over large areas. It takes time to bring the shut-down plants back on line, leaving us to sit in the cold and dark. We call this a black-out.

If demand out-strips supply, power companies must ration the available power to different users.
We call this a brown-out.

Human beings have not yet figured out how to store electrical energy very efficiently. We can't produce it and store it for later very efficiently nor economically. So, we use it as we produce it. Our nation's entire electrical power grid is based on the immediate use of what is produced, a delicate balance.

Of course the demand is always fluctuating. In the morning we turn on lights and make coffee, industry starts motors, offices turn on copy machines, mothers wash children's clothes - thus demand increases. At night businesses close, factories shut down, people turn off lights to go to sleep - thus demand decreases. The demand fluctuates by the moment during each day as well as by the long cycles of the seasons.

The reason our nation's electrical power system evolved into essentially one large interconnected grid is because that made it easier to balance the supply with the ever-changing demand. But this requires the ability to transmit electrical power over long distances. And we use a lot.

The United States consumed about 4,000 TWh (4,000,000,000,000 kWh) of electrical energy in 2013. Transmission costs seem to be about 2% of electricity's retail price. If we say 10 cents per kWh. Then it costs 0.2

cents per kWh for transmission. Therefore, it costs $8 billion to transmit 4,000 TWh. There are about 200,000 miles of high voltage transmission lines in the US. The wall's 2,000 mile line would be 1% of that. 1% of 4,000 TWh is 40 TWh which is $80 million transmission cost. If the transcontinental superconducting power lines in President Trump's Wall only handle half of that they would generate income of $40 million per year.

And so we have the challenge of balancing the grid's supply and demand requirements over large geographical areas as well as incorporating diverse intermittent power sources; wind, solar, etc.

Also we have the challenge of long transmission distances, generating power in one area and using it in another. The classic example being generating solar power in the remote unpopulated regions of the desert southwest to be used in distant population centers.

Attack on the grid can come from hostile powers in the form of electromagnetic air burst attacks, direct physical assault, etc. We are also susceptible to environmental attacks from sunspots, etc.

President Trump's Wall could provide protection and easy access for repair and upgrading of transmission lines. No need to bury or elevate them. The wall would provide; environmental protection, over-head cranes, lighting, etc.

Trump's Wall could provide a possible solution to most if not all of these urgent problems.

Solar Panel Power Generating Arrays

Life, through photosynthesis, uses only about 0.023% of all the solar energy that reaches the earth. Mankind controls only about 0.016% of the total amount of energy the Earth receives from the Sun. The Earth receives more energy from the Sun in an hour and a half then mankind uses in a year.

With regards to solar panel power generation, it is difficult to make calculations with confidence. There is so much seemingly inconsistent information on the internet. The state of the art is advancing so fast. Few are experts in the field. However, it seems that covering an area 2,000 miles long by 20 miles wide with solar panels would produce an amount of energy roughly equivalent to the entire U.S. energy consumption. That is energy from all sources; gasoline, oil, coal, natural gas, hydro, solar, everything. This gives a general guide to the order of magnitude we are considering.

Some have proposed attaching solar power generating panels to Trump's Wall. One of the main objections seems to be that the panels will not be properly oriented to the sun. Along with this is the assertion that solar panels on the wall could never generate enough electricity to pay for the wall. Fair enough. Both of these points seem to hold true.

A point most commentators fail to discuss is, if solar panels are installed on the wall how will the power be transmitted to the end-users? Ah ha, we already have our transcontinental superconducting power lines safely installed inside the wall.

So, if solar power generating panels won't work well on the wall, **don't put them there**.

Encourage private parties to install and operate solar panel power generating arrays on private land **next to** the wall. Why are there not already forests of solar arrays along our remote sunny southern border? Because they have no economical way of getting their product, (electricity) to market (population centers). Allowing solar arrays to hook-up to the transcontinental superconducting power lines in the wall would give them what they need. This would solve the above-mentioned objections and allow for much larger solar arrays.

President Trump's Wall needs private investment to reach its full potential. The calculation for a viable solar power generating array becomes much rosier when the costs of installing and maintaining transmission lines are removed. Of course individual investors would make their own calculations but this opens up income potential for thousands of adjacent landowners.

The practical details are very important but building is an inherently human activity. People build for human reasons to serve human needs and fulfill human desires.

When people become financially invested in President Trump's Wall (through jobs and ownership, rents and stocks) they will come to see their self-interest reflected in it and begin to see it as "their" wall, they will become emotionally invested, they will begin to care.

Plus, this could provide an excellent opportunity for people in Mexico to create jobs and make money. Offer to allow Mexican solar power generating arrays to hook into the wall's transcontinental superconducting power lines.

So, festooning President Trump's Wall with solar panels won't pay for it. Maybe even much larger adjacent solar arrays won't pay for it. That could be a problem, if we intended to stop there. But solar panels are only a small part of its much greater potential.

Desalination Plants

Why bother? Good heavens. Say we built two huge solar power generating arrays, each 10 miles wide down both sides of the 2,000 mile wall that produce as much energy as the total amount our entire country uses now. What would we do with the power?

Aside from the fact that solar arrays of this size are unlikely, if for no other reason than economics, we actually would have a use for power.

Desalination.

The American Desert Southwest is headed toward a water crisis. People are attracted to the region and cities are growing. This is putting a severe strain on the limited water resources of this arid environment.

> *"The demand for a reliable and secure supply of water for a growing region must be met by the carefully selected and economically efficient development of new water."*
> *- Kay Bailey Hutchison Desalination Plant*

As of 2015 the price of tap water in rainy Seattle, Washington was virtually the same as in arid San Diego, California, about 0.9 cents per gallon or about $400 per acre-foot. Without getting into the reasons for this seemingly counter-intuitive observation, it goes to show that, in our modern world, local climate has little to do with the price of water.

There has been much debate and ill feeling generated over different local's water rights and the long term viability of their water supplies.

So, more clean water is needed, it is already in demand.

Some advocate using less, the reduction of our standard of living, as if that were the moral high ground. Others insist on bullying and fighting over finite resources. But, if we choose, we can create abundance and in so doing make better lives for ourselves and our families. The issue is not water, the oceans are full of it. The issue is pure, clean water delivered to those who use it.

To create clean water, through natural or manmade means, requires energy.

The Sun's energy evaporates water off the oceans, transports it and precipitates it over land. We collect this for our use. But we have no control of this process of creation, the weather, and only some ability in building reservoirs to catch and store this bounty the Earth gives to us for free, through no effort of our own.

This has served mankind up until now but it is time for us to take control of this vital resource, clean water. We should strive to have the same control over clean water as we have over electricity. **We create electricity, we conjure it out of nothing.** We create electricity and deliver it to those who need it, when they need it. There is no reason we cannot or should not build this capability with clean water. We need to take control of our entire water use cycle. Create clean water on demand and deliver it where it is needed, when it is needed. At present we do a good job of delivering it, not so great at creating it.

For all our modern advancements, we are still held hostage to forces beyond our control, the weather. Deluge is as destructive as drought and there is precious little we can do about it.

We need to be able to create our own clean water. Since we cannot control the weather, desalination seems to be a primary option. There are several different desalination technologies from simple boiling to reverse osmoses (forcing saltwater through membranes) and even electrically driven shockwaves, to name just a few.
      http://news.mit.edu/2015/shockwave-process-desalination-water-1112

They each have their pros and cons. Boiling is very simple but takes more energy, osmosis takes less energy but has many moving parts and filters clog. A careful evaluation of all available technologies needs to be done to make an economical decision for each new facility.

There is the other issue of cost of water. As pointed out above, the cost of residential water is about $400 per acre-foot (an acre-foot is a volume of water one acre by one foot deep, 325,851 gallons).

Internet research seems to indicate that it takes between 14 and 20 kWh of energy to desalinate one thousand gallons of water, say 18.5 kWh. At 10 cents per kWh it costs $1.85. Which means ($1.85 x 325,851/1,000 = $602.82) the energy required to desalinate one acre-foot of water costs about $600.

However, total actual desalination costs seem to be between $2,300 and $5,000 per acre-foot.

Compare this to the $400 per acre-foot residential water cost above and we see a problem.

Desalination cost is high and it does not seem to come from the energy used (only 10%-25%), rather the physical plant; tanks, pipes, pumps, filters, etc.

Perhaps this indicates that the water purification system that relies more on energy and less on equipment would be where to look first, i.e. boiling water and collecting the steam. It seems worth looking into.

Is there a way around this? It is hard to say, but one thing seems certain, we don't have a choice. We either take control of this or it will forever control us.

To put things into perspective, assuming people in the U.S. each use 100 gallons of clean water per day. It would require about 200 TWh of energy per year to desalinate all of it. Total US electrical usage is about 4,000 TWh per year. So it would require about 5% of our electrical energy usage to make 100% of our residential clean water needs (residential clean water is about 10% of total fresh water usage; farming, industrial, etc.).

The bottom line is, the controlled creation of clean water takes energy, lots of it, but not more than we can handle.

Energy is the key. The ability to control the production and use of energy makes life better. Life is not a zero sum game, unless we chose it to be. We have brought up solar power generating panels but we will discuss energy more later.

But, after all this, what does President Trump's Wall have to do with making clean water?

Well, it is a ready-made pipeline corridor.

- As the wall extends from sea to sea, it is a natural fit to build a desalination plant at each end.

- Install pipelines in the wall's interior; secure, protected, inside with overhead cranes, power and lighting for ease of installation and maintenance.

- Convenient location to supplying the population centers of the American Southwest.

But most of all, lots of clean water will be required to serve the needs of industry and residence that will populate President Trump's Wall in the next step of its development.

Let us proceed.

In The Beginning Wrap-Up

So there is the first stage.

A transcontinental power transmission line which provides the missing component to make viable remotely located solar and wind power generating facilities which provide the energy to desalinate water which the American Southwest needs. This system would create many opportunities, businesses and jobs.

But how to actually build it?

Use railroad construction as an inspiration. Railroads are repetitious constructions of prefabricated parts; ties, rails, etc. These parts are mass-produced and then transported to the construction site for installation. The installation work is heavy and hard but it is simple and repetitious. This is very efficient. What is more, the track that is laid is used as the transportation system to bring more prefabricated components to the railhead to lay more rail ahead. Thus the railroad serves as the infrastructure to build the railroad. No separate transportation system, i.e. roads, etc, need be built.

Using this same construction model let us now consider how to build President Trump's Wall.

It has been proposed to build a wall with a high-speed road next to it on the American side, all within about 60 feet of width. This is fine and all very doable. But in the end we are left with a wall and a road.

But what if we consider building using the railroad construction idea of using the wall as the transportation system to bring prefabricated components to build the wall?

If the wall is conceived of as a box-beam, two parallel walls with an elevated roadway for a roof. The high-speed roadway for border security is the roof. Between the walls; inside and protected, with overhead cranes, lighting and power, becomes a conduit for transcontinental superconducting power lines and fresh water pipelines. Now Trump's Wall has a greater reason for being. Now it pays for itself, its construction and maintenance. Now it attracts private capital because it produces a return on investment.

On a side note. In the past decade or so there has been research and experimentation with a glass road surface. Glass that does not get slippery when wet and is tough enough to drive on. A glass road on the roof could act as a solar power generating panel. A really neat idea is mounting LED's under the glass thus the roadbed could be lit up. Wouldn't that look impressive from space! But I digress.

Of course I have been conveniently vague and lacking in details. But that is not the point. There are many ways to proceed and America has lots of smart determined folks who can design, plan and solve problems.

The point is ... President Trump's Wall is no longer a Wall.

Now that we have this stage in place and we are being productive and making money, it is time to move onto the next stage, The Big Jump!

The Big Jump

So, we now have a functioning project. It makes money. It provides border security. It has transcontinental superconducting power transmission lines. It has solar power generating panel arrays connected to it. It makes clean water from the oceans. It has pipelines bringing water to people.

This is pretty good. Much better than just a wall and a road. But it could be so much more.

At this point President Trump's Wall is a building 60 feet wide and 30 feet tall with a road on top and utilities inside. But it might soon become obsolete, outdated, old news. When it starts paying for itself, we might just take it for granted and want more; more jobs, more investment opportunities, more productivity.

We have successfully built this type of grand project before. The Erie Canal opened in 1825. World renowned, it was the technological marvel of its day.

Today the Erie Canal seems small and insignificant, a quaint and anachronistic tourist attraction, certainly not capable of fundamentally changing our nation's history. The decision to build it was a contentious, decades long fight. Many people thought it was crazy. It pushed the bounds of known technology. Its proposed construction cost was barely imaginable, amounting to one third of all banking and insurance capital in New York State.

But within 10 years of its completion it had paid for its construction, operation and maintenance and its continued income offset the New York State budget by 2/3's.
Can we learn from that? How did they do it?

How was it that this endeavor, our country's first major construction project, was so successfully managed? It stretched the technology and financial resources of our fledgling nation to the limit. Why did it not go bankrupt as so many projects had before?

Because, it was intentionally built in stages with each stage designed to create income. Because it allowed private businesses to operate on and around it. It even allowed private tributary, so called "feeder", canals to tie directly into it.

If we just look at the total construction time of 8 years it hides some very important aspects of the way the construction was managed that lead to the Erie Canal's ultimate success.

First proposed in the 1780's, construction did not start until 1817. It was a monumental task for its time, stretching 363 miles long with 50 locks and a 565 foot elevation change, including scaling the Niagara Escarpment's 80 vertical feet of limestone. The technical hurdles were daunting for its day, not the least of which was the Malaria of the swamps near Syracuse, NY.

The Erie Canal was an immediate success. It exceeded its designed capacity from the very start. Within 10 years it had paid for its construction, maintenance and operating costs.

But almost immediately people were frustrated with it, too small, too congested, too slow, and this was a good thing. People's frustration meant that they were using it, deriving value from it, making money using it. And everyone benefited, from the young "Hoggee" mule driver boys to wayside hoteliers to adjacent farmers to large corporations.

Initially it had been intentionally built in a minimalist way, not as a big beautiful perfect project. A mere 40 feet wide by 4 feet deep, with bridges only 4 feet above the canal water, "low bridge, everybody down". Just good enough to connect New York and Buffalo so products could flow, money could be made and most importantly so people could become invested in its success and survival. People came to care.

When construction began in 1817, it commenced in easy flat terrain so construction workers could learn and success could buoy public confidence. By 1820 the section between Utica and Syracuse was complete and commercial traffic started immediately. Thus in two years an income stream had been started. By 1823 the 250 miles from Brockport to Albany was opened and the added commercial traffic increased the income stream. When finally completed in 1825 it immediately exceeded its designed capacity.

Why? Because people had already been using it for 5 years and understood its potential. They had 5 years to prepare to grasp the economic opportunities the Erie Canal unleashed.

They saw their own self-interest in the Erie Canal. So, when improvements were proposed it was in response to peoples desires and investment capital was easy to find. Opened in 1825, by 1832, just 7 short years, work to enlarge the canal was begun. Now at 70 Feet wide and 7 feet deep with bigger locks, it could accommodate larger canal boats with less congestion. Faster transit times and bigger boats made transportation costs come down even more and goods became cheaper. The economy grew and more boats ran on the Erie Canal.

Had the builders not allowed any traffic until it was 100% completed, people would have had to figure out how to use it and learn what infrastructure to build; boats, docks, cargo canes, gang ways, hotels, stables, etc. As it was, new ideas were tried and conventions established as the canal expanded. No doubt many of these facilities were built along the canal at the same time as it was being lengthened. Thus, when a new section of canal was opened, the infrastructure was already in place and traffic flowed. And much of this infrastructure was built with private investment. The $7,000,000 canal attracted untold millions of private capital.

Furthermore, as a consequence of its construction sequencing and management, the Erie Canal's final cost was only a scant 2% over its original budget. When is the last time we heard that about a large government construction project!?

But, we are not building a canal, we are building a wall.

So, what is my point? My point is, because people want more money-making opportunities, we are about to "dig a bigger canal", a much bigger "canal". A bigger wall, a much bigger wall.

With that in mind, let us begin.

Shipping Corridor From Sea To Sea

The United States of America spans a continent and thus faces two major oceans giving it access to two of the world's major economic zones, Europe and Asia.

This is a great advantage but South America creates a very long shipping lane between our East and West Coasts. The Panama Canal is a huge boon in relieving this problem but we do not own it, we do not control it. We must rely upon others for what is vitally important to us.

Even with modern technology, transportation by water cannot be ignored. It is desired for its low cost and high efficiency. Seaborne transportation dominates worldwide trade comprising perhaps 80% of its volume and 70% of its value.

We Americans have built seaports on both coasts to handle the vast amount of commerce that flows into and out of our country. But seaports have the frustrating problem of becoming the victims of their own successes. As more commerce flows through them the cities around them grow. Thus the port facilities and the city, which need each other, find themselves competing for available space. What is more, the city chokes the vital transportation corridors, road and rail, that connect the port to the country's interior. Thus seaports find themselves locating industrial facilities on very expensive seaside real estate that has poor land transportation access.

Furthermore, being located on the ocean, seaports are exposed to the effects of weather. Historically, coastal topography has dictated seaport locations. A natural harbor protected from wind and wave being of paramount importance. But even with that, no port is completely immune to the effects of weather.

More than 90% of worldwide non-bulk trade goods and 25% of bulk goods (grains, etc.) are transported in shipping containers.

Container ships are classified by the number of twenty foot long shipping containers they can carry, referred to in units called Twenty Foot Equivalent Unit (TEU).
Worldwide there are about 161 ships of 10,000 TEU or greater and roughly 51 ports that can handle them, with more being built all the time.

Well, what does all this have to do with Trump's Wall?

Trump's wall, at this stage, has established a transcontinental corridor. Let us expand it to accommodate shipping.

This does several things:
- It moves seaport facilities inland thus limited seaside space is no longer an issue.
- It moves seaport facilities inland thus ground transportation through congested port cities is no longer an issue.
- It moves seaport facilities inland and indoors thus weather becomes almost a nonissue.
- We take control of our transcontinental shipping.

- Security and customs become much easier to control.

So what does this look like? How to proceed?

The proposed wall structure will have to be enlarged to accommodate shipping. This new wall will dwarf the existing 60 foot wide, 30 foot tall structure. It will need to accommodate multiple rail lines (for moving shipping containers), transfer stations, facilities (for manufacturing and commercial) and a Hyperloop.

It is more efficient and less expensive to transport cargo on larger ships. Larger ships use less fuel per pound of cargo. The push is toward bigger ships, much bigger. A large ship of the 1990's carried 5,000 standard 20 foot long shipping containers (TEU), today the largest ships carry in excess of 15,000 TEU. One of the main limiting factors is not how big we can build ships, but the seaports and waterways capable of handling them.

In 2016 the Panama Canal completed a new, bigger third set of locks in order to be able to handle the larger ships. This was a major, very expensive undertaking. There are few seaports that have the equipment needed to handle these huge ships. Some U.S. seaports have, or are in the process of building, larger facilities to handle them but it is very expensive so few can justify the expense.

Can we take advantage of this need? Can we serve it efficiently and economically?

Container ship facilities could be built in the oceans at each end of President Trump's Wall to offload and on-load containers from ships. The facilities would not need the large land area that seaports typically need because the containers would be immediately fed into the wall for handling in the wall's transit centers tens or hundreds of miles inland.

Shipping companies typically demand that ports offload their ships in 24 hours. This requires the ports to have larger container storage areas and increased road and rail capability. Trump's Wall would handle all of that.

Should even larger ships be introduced in the future, the infrastructure to handle them would likely be relatively inexpensive to upgrade in President Trump's Wall compared to conventional ports or canals because container storage areas, ground transportation and lock size would not be an issue.

In accordance with our national interest, each container entering the wall could pass through modern security and customs checks, a much more thorough system then at present.

We could use conventional rail lines laid inside Trump's Wall to transport cargo into and across our nation. Multiple rail lines, possibly in vertical tiers if needed, with shunts into transit centers could move vast quantities of shipping containers efficiently.

Shipping companies could be invited to build large transit centers attached to the U.S. side of the wall (Maersk, Hyundai, FedEx, Amazon, etc.). These transit centers, tied into the wall's transcontinental rail line, would thus be connected to both oceans as well as truck, train and air transportation to the U.S. interior. These transit centers would essentially be seaports. They would have the advantage of seaports, being connected to the oceans, without the disadvantages of limited land, restricted interior transportation corridors and weather problems.

For the transporting of shipping containers we could use electrically self-propelled railroad boogies designed to lock onto each end of each shipping container. A boogie is essentially train wheels on an axle, a two or four-wheeled railroad cart. These could each have their own electric motors that receive power from electrified track rails as modern subway trains do now.

With each shipping container acting as an independent self-propelled train car, the assembling of long trains would not be necessary.

Much as the post office sorts letters, the shipping containers could be tracked as they travel down Trump's Wall and be shunted into the appropriate large shipping centers located along the U.S. side of the wall. There, they could be unloaded and reloaded to be sent out again to ships in the Atlantic or Pacific Oceans or to truck, train or air transportation to the interior of America.

For some cargo, containers could be offloaded from ships in one ocean, transported and on-loaded to ships in the other ocean, thus obviating the need to transit the Panama Canal.

Let us not forget tankers. These ships carry liquids. Two of the main fluids are oil and Liquefied Natural Gas (LNG). We already have water pipelines in Trump's Wall so adding oil and gas pipelines would fit right in.

Furthermore, bringing oil in from subsea oil wells would allow oil refineries to be located along the wall.

Whereas container ships are rated by the number of containers they carry, tankers are rated by their weight. The largest "super-tankers" today are in excess of 250,000 tons. For a comparison, the battleship USS Missouri weighs about 60,000 tons when armed to the teeth.

You may have noticed in my letter to Dr. Carrick that I introduced the idea of lifting entire ships on to a rail system to be transported from one ocean to the other. Well, in my enthusiasm I failed to consider the realities of what I was proposing. It turns out that the heaviest trains seem to weigh about 4 tons per lineal foot. The typical train being 13 feet wide. Compare that to a 200 foot wide container ship weighing about 125 tons per lineal foot or tanker weighing about 400 tons per lineal foot and my enthusiasm begins to look ridiculous, much to my chagrin. And Mag-Lev trains are of no help here as the heaviest seem to weigh only about 0.4 tons per lineal foot. Even if we adjust for width, ships are still about 2 to 6 times heavier per lineal foot than the heaviest trains.

It seems train weight is limited by bridges and railroad grade (steepness) so it may be possible, but unclear, if the rails, wheels and suspension could take this greater load. Of course an empty ship would be much lighter so who knows... That is until one realizes that container ships are routinely in excess of 200 feet tall. So, there would have to be a strong reason to do this ... but it would be an impressive sight. And hey, maybe one day we will even build ships in the wall. Hmm. But, Moving on.

This integrated system invites innovation and automation for the rapid and efficient movement of cargo, oil and gas into and out of our country. And it gives us control of a commercial and strategic shipping route vital to our national interests.

Well, this is great. It gives the United States some important capabilities we lack at present. But is this all? No. There is more. We are pursuing a greater vision. Let us continue.

Manufacturing - Commercial - Residential

If President Trump's Wall is thought of as framework that houses a transportation/material handling corridor that moves shipping containers to and from large transit centers and one realizes that America imports more than it exports, then naturally we end up with a lot of empty shipping containers. What to do with them?

They could be stored in a bank of standard sized receptacles (like a bank of post office boxes) built into the wall's interior. But can they not be put to use? Must we just store them?

As it happens there is already an entire industry of companies that design and build interiors for shipping containers for different purposes; offices, shops, even homes.

But we need not stop there. We could think of bigger "boxes". Much bigger standard size "boxes" say, a cube 200 feet on a side, basically a 20-story building. Thus we would have two standard sized "boxes", one "large" and one "small". The interior of these "large" "boxes" could be designed and built to suit individual needs for; manufacturing, industrial, residential, office, entertainment, etc.

Receptacles/cubby holes could be built into the interior of Trump's Wall to receive the large standard size "boxes".

To move a "box", it would be drawn out onto the transportation corridor, moved down the wall and inserted into a vacant receptacle.

Some of the large standard size "boxes" could themselves be made to be frameworks with "post office box like" receptacles that receive small standard sized "boxes", i.e. shipping containers. This would be similar to an apartment building full of shipping containers, shipping containers slotted into receptacles on each floor. These shipping containers could have interiors personalized to the customer's/owner's needs.

Thus a person or small business could own their own home or office and move it at will anywhere along the wall.

Using shipping containers as the small standard size "box" might make sense. There are a lot of them, roughly 17 million worldwide, and a lot of them are sitting empty in our seaports.

https://www.uh.edu/engines/epi2879.htm

Designing and incorporating standard quick-connect connections for water, drain, ventilation, communications and electrical power throughout the wall and in each standard size "box" would make moving inexpensive and quick.

The ability for any home or business to easily relocate elsewhere along the wall would give people options.

With an interchangeable standardized system such as described, any home, business or factory could be easily moved elsewhere in the wall. Assuming each section of the wall was considered part of the jurisdiction it passed through, the wall would be a long collection of different local governments. With this array of choice, it would be difficult for people and companies to be overly regulated or overly taxed by any one jurisdiction because they could just move, lock, stock and barrel, if they were unhappy. This would force local governments to compete for residents and businesses, thus limiting the power of government and increasing personal freedoms.

So, with residences and businesses spread for many hundreds of miles along the wall, how do people travel between them? How do we get to work everyday? This need is what we will cover next.

Hyperloop

So, we have a structure with large shipping centers, manufacturing facilities, commercial offices, and residences located along its length. It also houses power lines, and pipelines that need to be serviced and regularly upgraded.

But these are scattered over a length of 2,000 miles. Being stretched in a long line makes access awkward. Is there a way to efficiently use this layout? Is there a way to "bring things together"? Can we make the distances shorter?

Well, No and Yes.

No. Because the long thin layout is a requirement of President Trump's Wall, which is beyond our control.

Yes. Because, whereas we cannot shorten the distances between things, we can shorten the travel time between them.

The idea of the Hyperloop was introduced several years ago and brought to popular attention by Elon Musk. It is a high speed transportation system.
     https://en.wikipedia.org/wiki/Hyperloop

This idea relies on the realization that, as the speed of an object increases, the drag force on it (from the air in this case) increases rapidly out of proportion and therefore greater and greater power is required to overcome it. Thus at high speeds a great increase of thrust is required to increase the speed only a little.

Drag force is produced because as an object moves through air it must push air molecules out of its way. Boats do this in water, which we see represented by the wakes they create.

The common way of dealing with drag is to make the object's shape more aerodynamic. This reduces the drag, which reduces the power requirements. We see this in the sleek forms of race cars and the funny helmets of bicycle racers.

But there is another way... get rid of the air. There is no aerodynamic drag on an object moving through a vacuum because there is no air. There are no air molecules to push out of the way.

Moving without aerodynamic resistance allows much higher speeds. Not only does it require much less power but things don't get hot. Just moving through the air fast enough can make objects blazingly hot. We saw this tragically on February 01, 2003 when, upon reentering the atmosphere with damaged heat shielding, the Space Shuttle Columbia was destroyed by 3,000 degree Fahrenheit temperatures caused by air friction.

So the Hyperloop relies on objects moving through a near vacuum, cars moving through a tube that has had most of the air pumped out of it.

The design calls for car speeds of around 700 miles per hour. At 700 miles per hour, a daily commute to work of 1,000 miles would not be out of the question.

The Hyperloop is a neat idea. But one of its biggest challenges, after deciding which locations to connect, is getting a right-of-way between those places. Emotionally charged battles over viable corridors are very contentious and have scrapped more than one proposal.

Then there is another issue. Elevated or tunnel, which to choose. There are proponents of both but neither option has moved projects past stalemate.

Security is intertwined with all of these aforementioned considerations. Protecting a long thin structure like a Hyperloop tube requires a huge amount of manpower and is almost impossible. Security is a fact of our modern world and cannot be ignored.

Objections to the Hyperloop are that it would be too vulnerable to power outages and terrorist attacks. Built as proposed this is possibly true, but not so much if located within the protective envelope of Trump's Wall.

The combination of Trump's Wall and the Hyperloop would create an excellent synergistic relationship.

The Hyperloop needs a corridor, security, power, cranes for construction and easy access for maintenance. Trump's Wall can provide all of these.

Conversely, Trump's Wall needs a reliable, convenient, high-speed transportation system to realize its full potential. A Hyperloop can provide this.

Most importantly Trump's Wall gives the Hyperloop a reason for being.

In the greater worldwide view of transportation alternatives, it is true the Hyperloop is often a good solution for distances too long for cars and too short for planes, but few see it as necessary. As transportation for these middle distances, it is just another option. It tries to fill the place between cars and planes, but it is seen as an expense, a "nice to have". It must justify itself on the income from passengers. Passengers whose travel needs are already being met by other means.

Most of the public conversation about the Hyperloop is not about the construction technical details, as important as they are, but about what locations to connect and how to get a right-of-way.

Not so with President Trump's Wall.

Trump's Wall needs it and it needs the wall. The Hyperloop has a greater reason for being when located in the wall. It makes the wall viable, without it the wall's potential is greatly reduced.

Without a high-speed transportation system security, emergency and maintenance services would be less responsive. People would have less choice where they may live.

With the ability for people to, at will, reliably transit its entire length in three hours, President Trump's Wall would become a different thing. It would function differently. It would open possibilities that we haven't thought of yet.

There is much activity at present in the development of technology for the Hyperloop. Installing it in Trump's Wall would call for even more. One Hyperloop tube would not be enough. There would need to be several; express tubes, commuter tubes, etc. They need not be laid side-by-side like train tracks. Multiple tubes could be stacked in clusters and/or located at different levels throughout the wall.

The Hyperloop has another interesting possibility. If each car is independently self-propelled and they are not assembled into trains, then more autonomy is possible. It may be that people will buy or lease their own cars. Equipped with internet access one could be productive in transit. Some may even make their Hyperloop cars moving offices, which dock into their employer's office building for work and return them home at night.

An interesting thought is to use shipping containers for Hyperloop cars. Design Hyperloop "end caps" that lock onto the ends of standard sized shipping containers, the inside designed and built to whatever is desired. Thus we may find similarities with the privately owned packet boats of the Eric Canal in that innovative design allowed them to be reconfigured for relaxation, meals and sleeping. But it remains to be seen if this is a practicable idea.

There are many exciting possibilities that capable, energetic people are already experimenting with. All-important standards will eventually be settled on and development will continue.

But the bottom line is President Trump's Wall needs a high-speed transportation system. It need not be a Hyperloop, but it needs to be something.

The Big Jump Wrap-Up

The Big Jump is BIG.

So what would this look like?

This addition to an existing 60 foot wide, 30 foot tall structure would dwarf it. It would need to accommodate real estate, a Hyperloop, factories and even an entire railroad system.

If the primary real estate unit is a 200 foot cube then the real estate zone, along the inside of the northern wall, would be some 250 to 300 feet wide. The transportation corridor along the inside of the southern wall would need to be perhaps 700 feet wide. Thus a structure 1,000 feet wide and in excess of 200 feet tall might be expected.

But for such a mammoth undertaking, how to build it, how to finance it?

Build it in stages.

It is all in how we conceptualize this project. There is no absolute necessity for this entire expansion to be built at the same time, as a unit. It need not be built coast to coast all at once. It could be built in sections with money made at each stage of development.

The shipping container transportation system could begin with transit centers built just a few tens or hundreds of miles inland from each coast. As demand increases and income streams are created, the wall could be expanded in stages until finally a transcontinental rail connection is made.

And let us remember, this has been done successfully before. The Erie Canal was built using a staged construction process. We can take inspiration from those who have gone before us.

A Hyperloop, with minimal capacity, would serve utility systems maintenance and security needs from the start. This could be expanded based on demand as more manufacturing, commercial and residential units were brought on line.

As for how to proceed with construction, never interrupt the services in the original wall. It is of primary importance to preserve existing income streams. First build the shipping container transportation rail lines beside the original wall and create more income. Then expand the existing structure to enclose the railroad and add a Hyperloop, increasing Hyperloop capacity in accordance with increasing demand, and so on.

It may be that people will come to live and work in President Trump's Wall.

To some this idea may seem overwhelming, foolish and vain.

The question is Why? Why build such an audacious thing in the first place? Don't we have enough electricity and shipping and factories now? Well, yes we do. And if the purpose of this proposed President Trump's Wall were merely to replace existing capabilities it would indeed be a fools errand. But it isn't. It serves a much grander vision.

Everything to this point is just laying the foundation for the grand purpose of this proposal to be built upon, to create an integrated whole for a greater purpose.

To go to space, not as a visitor but to work. Manufacturing in space on an industrial scale, made possible by low-cost access to Earth orbit, will change the course of human history and declare mankind a spacefaring race.

With that in mind, let us proceed to the final stage of President Trump's Wall, The Leap To Space.

The Leap To Space

The Big Jump was big. At this point we have proposed the audacious concept of gigantic inland seaports, made possible by our continent spanning robotized shipping container handling and distribution system. The scale of the proposed building called Trump's Wall is mammoth.

Impressive, but is this all there is? Did we set out to just build a big building? Is there no greater purpose? If not, this wall will never be built.

That which has been proposed so far has no abiding emotional appeal. Nothing to grab the imagination nor stir the human soul.

President Trump's Wall needs a reason for being, to serve a greater purpose, to embody a grand vision. To inspire a generation and so unleash human ambition and focus human potential.

Its parts need a grand unifying purpose that make them function as a greater whole. To concentrate human endeavor in a way that directs the path of human advancement. To change the very way mankind sees itself.

And so we arrive at our final section, The Leap to Space.

Space Launch System

Ah, to space. But what does that mean, "go to space"? Go where? The Moon? Mars? Why? What is the reason?

What we really want is access to Earth orbit. Regular, reliable, economical access to Earth orbit is the real prize.

Why? Because it would enable manufacturing in space, which would advance human understanding. It would be perhaps the greatest wealth creator in human history and it would change how mankind views itself.

Consider the possibilities afforded by manufacturing in the completely different environment of space. Super-strong amorphous glass-like materials, super hard metallic and ceramic coatings, perfectly grown crystals of large size, and many other materials with amazing properties, like superconductivity, that have been theorized but never actually made.

Building infrastructure in Earth orbit would create a staging area to fit-out and launch a great variety of expeditions to more remote destinations. And with this capability mankind could justly consider itself a spacefaring race.

But getting to Earth orbit is the first, and biggest, step in getting anywhere else. It takes as much energy to get from the Earth's surface to Earth orbit as it does to get from Earth orbit to the surface of Mars, 1.3 times more than to get from Earth orbit to the surface of the Moon, twice as much as required to get from Earth orbit to near-Earth asteroids. It takes a lot of energy to escape from Earth's gravity.

Because reaching Earth orbit requires so much energy, let us briefly consider the reality of rockets, or as Don Pettit says "the tyranny of the rocket equation".
    https://www.nasa.gov/mission_pages/station/expeditions/exped ition30/tryanny.html

The physical laws of motion, as quantified by Isaac Newton's mathematics, require typical manned rockets to be about 85% fuel, by mass, with only 4% of their total weight being payload (i.e. cargo). This roughly 4% payload figure is due to the merciless realities of physics. It applied equally to the Saturn V rocket of the 1960's as as it does to the newest SpaceX rockets of today, even with the intervening advancements in materials science, computing power and technology in general. Only 1% of the Space Shuttle's total weight was payload, because its reusable part weighed 3%.

Physics is a serious limitation to the access of Earth orbit.

An interesting side note. If Earth were 50% larger than it is, it would be physically impossible to escape Earth's gravity using the chemical propellant rocket technology of today.

As pointed out above, regular, reliable and economical access to Earth orbit would be highly beneficial. But, in order to realize this potential, we not only need to get there, we also need to be able to move enough mass to matter. Remember, we are going to space to work, to be productive, to make money. And manufacturing, with all its machinery and raw materials, is heavy.

With a payload limit of 4% of total rocket mass, a lot of mass means a lot of rocket launches. Since 1957 there have never been more than 150 rocket launches per year worldwide. Rocket launches over the past 20 years have varied from 50 to 80 per year.
    https://en.wikipedia.org/wiki/Timeline_of_spaceflight

Launch failures over the past 40 years seem to fairly consistently amount to about 5% of the total each year. Interestingly, the failure rate has not reduced with improved technology. This is likely due to the necessity of designing rockets with very small safety margins, in order to achieve weight reduction, which creates conditions where every small variation and transient load may cause catastrophic failure.

Modern rockets seem to carry about 10 tons of payload with the heaviest lifting about 20 tons. The least expensive cost of lifting payload into Earth orbit at present seems to be about $1,800 per pound.

We could expect to lift about 2,000 tons of payload into Earth orbit each year using chemically propelled rockets. (100 rocket launches x 20 tons of payload = 2,000 tons). Doubling the number of launches per year gives 4,000 tons. This may seem like a lot until one considers that one modest sized CNC milling station weighs over 10 tons.

Chemical propellant rocket technology does not seem to be a practicable solution to facilitate industrial scale manufacturing in space.

As mankind is not capable of violating the laws of physics, to overcome this limitation we must use different technology. We must find a different way.

As it happens there is another paradigm for access to space. One that people have been working on for over 20 years. It is called StarTram space launch system.
https://en.wikipedia.org/wiki/StarTram

James Powell dreamed up the concept of magnetic levitation and after several years of development work received a patent for it in 1968. So-called maglev is used today for high-speed trains. In 2001 he proposed using his maglev technology to launch rockets.

Chemical propellant rockets must carry their fuel with them, which is heavy. But another way to approach this problem is to leave the rocket's power source on the ground and shoot the rocket into space. This way, instead of the rocket's mass being 85% fuel it becomes perhaps 85% cargo.

The StarTram is similar to the Hyperloop in that it would propel its rockets down an evacuated launch tube to eliminate wind resistance. StarTram would use ground based power plants to propel maglev rockets into space. Its 900 mile long launch tube would have the final 150 miles or so of it levitated so that the end would be roughly 14 miles high. At 14 miles of altitude it would be above about 95% of the Earth's atmosphere so when a rocket left the launch tube there would be very minimal aerodynamic drag.

How to levitate to 14 miles? There is an interesting electromagnetic effect. When current flows through a wire it induces a magnetic field around it. Two parallel wires carrying current are either pulled together or pushed apart depending on how their magnetic fields interact. Thus, a cable on the ground repelling a cable on the launch tube festooned with guy wires to keep it in place and Voila.

To use this levitation method over these distances would require lots of power and very high current, perhaps 200 million amperes. But remember we have lots of power generation in Trump's Wall.

Not only would StarTram and Trump's Wall be a perfect fit, each providing what the other needs in order to justify themselves, but the roof of a 2,000 mile long "wall" is a perfect location for a 900 mile long launch tube.

StarTram is designed to deliver one 70-ton payload to Earth orbit per hour. That is about 600,000 tons per year. Roughly 150 times more than our rocket scenario above. To match this would require 30,000 rocket launches per year, more than 80 per day.

> https://www.space.com/38384-could-startram-revolutionize-space-travel.html

StarTram proposes to deliver cargo to Earth orbit for about $20 per pound which is only 1% of the present cost. Even $100 per pound would be only 6% of the $1,800 charged for delivery by chemical propellant rocket.

So, 150 times more mass delivered to Earth orbit for 1% of the cost. Perhaps this technology is worth looking into.

In 2002, the dollar cost per ton mile (the cost to move 1 ton 1 mile) for the four primary transportation modes was:

$ per ton mile
Air $4.63
Truck $0.37
Rail $0.03
Ship $0.01
http://richardtorian.blogspot.com/2012/01/cost-per-ton-mile-for-four-shipping.html

This shows costs for each mode to be separated by roughly a factor of ten (except the shipping to rail factor of 3).

Low Earth Orbit is defined as an altitude of 1,200 miles or less. Geosynchronous orbit is an altitude of 22,236 miles.

If we say that orbit is at an altitude of 1,000 miles, and we convert our table to show the cost of moving one pound 1,000 miles, then we enable a comparison with StarTram.

$ per pound 1,000 miles
StarTram $20.00
Air $2.32
Truck $0.185
Rail $0.015
Ship $0.005

Interestingly this maintains the ratio, showing StarTram to be roughly a factor of ten increase above air transportation.
https://www.space.com/38384-could-startram-revolutionize-space-travel.html

StarTram potentially represents a 150 times increase in mass of cargo moved and a 99% price reduction. This is very similar to the magnitude and effect the Erie Canal had on transportation in its day.

Pretty neat but pretty expensive, no doubt. Construction estimates for StarTram are roughly $80-$100 billion. Holy Cow!

Ah, but consider. Launching 600,000 tons (1,200,000,000 pounds) per year at $20 per pound (1,200,000,000 x $20) is $24 billion per year gross income. Hmm, perhaps there is hope for it yet.

I have described StarTram above, because I think it is cool, but it is only one of several proposed non-rocket space launch ideas.

https://en.wikipedia.org/wiki/Non-rocket_spacelaunch

They fall into four major categories:

- Static Structures:
    Tower
- Tensile Structures:
    Skyhook
    Space Elevator
    Endo-Atmospheric Tethers
- Dynamic Structures:
    Space Fountain
    Orbital Ring
    Launch Loop
    Pneumatic Freestanding Tower

- Projectile Launchers:
  Electro Magnetic Acceleration
  - Mass Driver
  - StarTram
  Chemical
  - Space Gun
  - Ram Accelerator
  - Blast Wave Accelerator
  Mechanical
  - Slingatron
  Air Launch
  Spaceplanes

There is a modern perception held by some that the Erie Canal was not worth building because it was soon displaced by railroads.

Opened in 1825, by 1842 (a scant 18 years later) a railroad extended its entire length. The railroad's faster speed soon captured most of the passenger traffic from the canal. However, in 1852 the Erie Canal still moved thirteen times more freight tonnage than all the railroads in New York State combined.

Augmenting its other positive effects, it is undeniable that by a purely financial measure the Erie Canal was a fantastic success.

Moreover, one must consider that the Erie Canal created the conditions that demanded the railroad. Not only did the canal move people to our nation's interior, greatly increasing its productivity and thus the demand for transportation, but it very likely was used to build, at least parts of, the railroad.

Every society, every technology and every person goes through a process of development. The development process may be streamlined and compressed but no steps can be skipped over.

It is reasonable to assume a similar life-arc of development, use and obsolescence for a space launch system where it will help build and create the demand for the very thing that will replace it. Where it to will one day become just a curiosity.

There is a fascinating idea of a Space Elevator. This is a cable anchored to Earth and extending to a counterweight beyond geostationary orbit in space. Geostationary orbit being 22,236 miles above Earth. The cable would be held in-place by the tension created by centrifugal forces (imagine the tension created when spinning a weighted string around in circles). Vehicles would climb up the cable to space and back down.
     https://en.wikipedia.org/wiki/Space_elevator

It is true that this cable would have to be stronger and lighter than any material known to mankind. Carbon nanotubes have potential but no one has actually made any length of cable that meets the criteria let alone figured out how to manufacture tens of thousands of miles of it.

But that is okay. There will be time. A StarTram, or some other space launch system, will provide orbit access and create the demand for space elevators. With that kind of incentive it is certain to have many bright enterprising folks racing to solve the problem. It may be that such a cable could only be manufactured in the ultra-dry, ultra-clean, weightless, vacuum of space. It may be that the need will be met by entirely different means. Who knows what the future holds.

But somebody must have some idea how to make such a cable because Obayashi corporation has plans to have its first space elevator operational by 2025 (just 7 years from now).

> https://www.quora.com/How-plausible-is-the-idea-of-a-space-elevator

But, hold on a minute. An electrically powered space launch system. Just exactly how much electricity will that require? Where will it come from?

A very good question. This is what we will cover next.

Nuclear Power

There are only four sources of energy known to human beings.

1) Energy from the inner Earth cooling, i.e. Geothermal.
2) Energy from the Sun. This includes high-energy particles, electromagnetic radiation (light), wind, wave, fossil fuels, biofuels, muscle power, etc.
3) Lunar. This is manifest in tides.
4) Nuclear.

To benefit from three of these four energy sources we have only to access it or start a fire. Only nuclear has demanded from mankind understanding and purposeful development in order to posses it.

It seems the energy requirements for a Star Tram launch system is expected to be 50,000 - 100,000 Megawatt electric (MWe) (we will assume 100,000 MWe). This is roughly 10% of the present total United States electrical energy generation capacity.

In order to maximize the space launch system we really need an on-demand power supply. Our plan for Trump's Wall already has solar panel power generating arrays but it doesn't generate power on demand, because it gets dark at night. Some other power generation source needs to be considered.

Our American psyche has been damaged in regard to nuclear power. The March 28, 1979 accident at Three Mile

Island Nuclear Generating Facility caused public concern, as well it should have. But the concern was pushed by some to the level of an irrational fear. This fear was reinforced and exacerbated by the subsequent high-profile nuclear facility accidents at Chernobyl of April 26, 1986 and Fukushima of March 11, 2011.

The primary result of the ensuing cascade of federal government regulation has been to greatly inhibit meaningful development of nuclear power generation in the United States. To be fair, there is ongoing effort in developing new and interesting reactor designs, so called Generation IV, but there has not been a new nuclear power plant ordered in the U.S. since 1978.

Can nuclear energy be dangerous? Certainly. But so can many things in life. Its risks can be safely managed if treated with respect, care and vigilance.

Nuclear power generation has some unique aspects that must be addressed, not the least of which is maintaining control of nuclear fuel so it does not get diverted to weaponry. Thus, the transportation, use and storage of fissionable material has expensive regulatory, logistical and security considerations that cannot be ignored.

Another aspect is cost. Existing nuclear power facilities are large. They typically exceed 2,000 MWe capacities, cost tens of billions of dollars and take a decade or more to build. This does not fit with our construction philosophy for President Trump's Wall, that of building in stages and creating income streams quickly.

But all is not lost. There are some innovative solutions actively being worked on. One is the idea of small modular nuclear reactors.

For example, NuScale Power, an innovative company based in Oregon, is actively pursuing the development of a factory made, self-contained, sealed nuclear reactor power module.
    http://www.nuscalepower.com/our-technology/technology-overview

This promising idea proposes to construct small modular self-contained, sealed nuclear reactors in a factory setting which would then be delivered to power generation facilities. This is an intriguing idea because it fits neatly with our Trump's Wall construction philosophy.

Small factory made units means low cost and fast construction. Thus an income stream could be created quickly and cheaply and capacity could be expanded by adding more modules to meet increasing demand.

And NuScale Power is but one example of ongoing innovation in reactor fuels, design and construction being actively pursued by different companies today.

I propose locating nuclear power generating facilities in President Trump's Wall.

Wait a minute! Are you proposing putting nuclear reactors in a building? A building with people in it? Well, yes I am. Of course the devil is in the details of design and construction but our Navy has operated nuclear powered ships and submarines with people in them for decades.

Multiple excavations could be made under the wall, one for each power generating facility.

http://www.nuscalepower.com/smr-benefits/safe

They would have everything they need, water, power, superconducting power transmission lines, maintenance facilities, overhead cranes, security, etc.

NuScale Power's small modular nuclear reactors are each 50 MWe with the ability for 12 modules to be operated from a single control room. Thus 600 MWe per NuScale installation (12 power modules).

Recall that a Star Tram space launch system requires about 100,000 MWe. 100,000MWe/600MWe = 167 units required to meet Star Tram space launch energy demand.
If we say, 180 units / 1,800 mile long wall = 10 miles.

One 600MWe nuclear power generating facility every 10 miles located in reinforced basement excavations under Trump's Wall seems very manageable.

Over the past few decades, a great deal of thought and effort has gone into designing adequate protections against potential failure from earthquakes, physical attack, etc. And let us remember these nuclear reactors will not be from the 1970's. Several of the newer reactor designs are internally stable in that they are not able to run-away and meltdown like Three Mile Island. Under a failure mode they shut down with no outside control or power required.

But let us take the benefits of cost, time and security derived from modular construction even a step further. Actually fabricate them in the wall. Build factories inside Trump's Wall to manufacture modular nuclear power reactors.

If modular reactors were manufactured in the wall and then deployed in it, many of the hurdles of fissionable material would be addressed. Trump's Wall would provide security for the entire reactor life cycle. Reactors could be manufactured, transported and used to generate power inside the wall. Old reactors could be updated to new generation reactors relatively easily using the heavy lift and transportation capability built into the wall. Reactors could live their entire lives in a controlled environment protected from the elements. Sensitive technology and nuclear fuel need never leave the secure environment of Trump's Wall. All power generating facilities would easily tie into the transcontinental power transmission lines already in the wall.

More than one hundred different nuclear reactor designs have been dreamed up by innovative minds over the past several decades. So, which to choose? Let the market decide. Many smaller sized reactors distributed within President Trump's Wall will allow for initial capital costs to be much lower and construction times to be much shorter. This will encourage many firms and different designs to compete.

Now there is the white elephant in the room, government regulation. Who can know if there is the will or the ability to create a regulatory environment that would allow, let alone encourage, putting nuclear power generating facilities in President Trump's Wall. But there is another option. Build nuclear power generating facilities in Mexico.

Oh, with this we can hear the proverbial pin drop. But wait. If reactors are build with U.S. know-how by U.S. companies and, with appropriate security agreements, deployed in Mexico it could work. Here is the idea. Build subterranean nuclear power generating facilities in Mexico attached to the side of Trump's Wall. Access between the wall and the nuclear facilities could be allowed enabling them to tie into the wall's power transmission lines.

Being in Mexico, U.S. nuclear regulations would not apply, thus reasonable agreements could be made. It is important to note that the companies involved will not be hamstrung by excessive U.S. government regulation but they would still be held responsible for their activities through the existing U.S. legal system.

The point is not to start an international incident. The point is to let our government know that standing in the way of progress comes with consequences and they have a choice to make.

With the United States providing design, engineering, manufacturing, management, operation, maintenance, investment capital, etc, and Mexico given the opportunity to supply land, construction, operation, maintenance and investment capital, it becomes a Win/Win solution that would produce jobs and wealth in both the U.S. and Mexico.

Excellent. Now we have a reliable source of power for the considerable demand of the space launch system. With that in mind, let us move on to why we are doing this in the first place. Manufacturing in space.

Manufacturing In Space

Now, finally, we, the human species, are ready to take the next step in our development. To begin the extraordinary adventure of spacefaring. And we must. For, to step into the unknown is the hallmark of life being lived.

Mostly our journey is about **Energy**. Developing our abilities to generate, control and use vast amounts of energy.

With substantial nuclear and solar power generation capacity connected by transcontinental power transmission lines, President Trump's Wall is the hub of a worldwide transportation system that handles materials and products. It is an integrated system that performs manufacturing and assembly operations then manufactures and loads space vehicles for launch into orbit. This massive flexible, reconfigurable assembly line is tied into a space launch system that regularly, reliably and economically gives access to the unique manufacturing environment of space, from which futuristic materials and products are returned to Earth.

Human beings are an adaptive species. We can live off of many different food sources, have adapted to different climates and as such have successfully populated most of the surface of our planet. But countless species have been unable to adapt to an environmental change and have gone extinct over the course of Earth's history.

Even today there are species that have very limited adaptability. Panda Bears, for example, can only eat certain leaves. We consider such species to be vulnerable and weak.

But if we expand our view to include, not the universe, not the galaxy but just our solar system then it becomes apparent that ... **we are the Panda Bears**. From this larger perspective we are very unadaptive.

Compared to our solar system we require a very limited range of pressure, humidity, gasses, temperature, gravity, visible light, protection from radiation, to name just some of the environmental factors we require for life.

The reality of the limitations of the human body are very real. We are biological creatures. Chemical cascades continually occur throughout our bodies. Everything from pressure, temperature, gravity and light, to name just a few, have immediate, long term and largely misunderstood and under-appreciated effects on them. Earth protects us. Space poses a very real threat to our physiology.

Recent research has shown that the human brain is exquisitely sensitive to cosmic radiation and high-energy particles from the Sun which cause cognitive impairment. Under stressful situations the executive functions of decision making and memory are adversely effected which are likely to last indefinitely.

"Our data provide additional evidence that deep space travel poses a real and unique threat to the integrity of neural circuits in the brain."
    - P. K. Vipan, et all
        https://www.nature.com/articles/srep34774

So, if human life requires a very specific environment are we stuck? No. We can extend our will into a wide variety of hostile environments through the use of remotely operated vehicles and robots. The realization is that we need not go in order to explore, we can be productive where no human can survive.

Now, there is another interesting way of looking at hostile environments. A ceramic coffee cup. A ceramic coffee cup embodies energy. It takes a lot of energy to fire the clay and even more to fire the glaze. The inside of a 2,000 degree Fahrenheit kiln is a very hostile environment to human life. Yet, by using this hostile environment to fire our ceramic coffee cup, we create an object useful in our environment. We have used the hostile environment to our benefit without going there. We have, in a sense, brought some of the hostile environment safely into our environment for our useful benefit.

Does this mean that no one need go to space? Probably not. Most likely people will need to go. But much like deep-sea divers they will not stay long. Offshore workers typically work month-on-month-off schedules. With a regular, reliable, economical space launch system, shifts of workers could be rotated through jobs in orbit. A schedule that minimizes the negative effects of micro-gravity, etc could be established.

Bear in mind that all orbital manufacturing will occur below the Van Allen Belts surrounding the Earth. The Van Allen Belts protect Earth by deflecting cosmic rays and high-energy particles from the Sun. So space workers will have some protection from this deadly radiation. Not to worry, there is plenty of room for manufacturing facilities beneath them as the belts start about 1,200 miles above the Earth's surface.

The environment of space is completely foreign to us. It presents many challenges but it also holds great potential.

Will we mine asteroids? Quite possibly, someday. Will we go to other planets? Possibly, someday. But these are not reasons to go to space today.

Manufacturing. Not as a replacement to the manufacturing we presently do on Earth, but as an adjunct to it. To make things we cannot on Earth.
    https://en.wikipedia.org/wiki/Space_manufacturing

What will these futuristic things be? Super-strong amorphous glass-like materials, super hard metallic and ceramic coatings, perfectly grown crystals of large size, and many other materials with amazing properties, perhaps like superconductivity, that have been theorized but never actually made.

Manufacturing in space could serve the needs of the small, computer chips to medical devices, and the very large, large single crystal turbine blades, solar panels and solar sails to name just a few possibilities.

Advancements in a host of areas would be made possible by the environmental characteristics of space:

Materials Science
Manufacturing Processes
Diffusion and Purification Processes

Of course, once economical access to space is in place, it seems logical to generate power there. Huge solar panel arrays seem obvious but there is also the intriguing possibility of harvesting power from the Van Allen Belts that encircle Earth.
https://wiki.energyxprize.heroxcontent.com/Energy_generators_sourcing_energy_from_Van_Allen_Belt

If the United States is to be the world leader based on a strong economy, the most powerful military and the most advanced technologies, then we need to dominate space. We need more than just satellites in orbit, we need an actual presence in the form of manufacturing and power generation.

Someday we may choose to go farther than Earth orbit. If we want a permanent base on the Moon we are not going to get it by using chemical propellant rocket technology. It just can't move enough mass into orbit fast enough nor cheaply enough. The future will require creative new thinking.

The Leap To Space Wrap-Up

So there we have it. Trump's Wall can be an entire system, a whole greater than the sum of its parts. A worldwide transportation system hub and integrated terrestrial manufacturing complex tied to an Earth orbit manufacturing infrastructure by a space launch system.

Without this transportation and manufacturing system supplying it, we may never be able to create and feed cargo laden space vehicles to a space launcher fast enough to realize its full potential.

Most of the excellent technological ideas proposed in this book need Trump's Wall. They are projects that have not been build because they have not been able to justify their existence. To build a space launch system in a remote location like Antarctica for the purpose of sending tourists on sightseeing rides to 70,000 feet is not a strong reason and would never have enough custom to pay for itself.

These projects have lacked three important things:

- Land - Lack of required right-of-ways
- Justification to exist - each is needed to create the whole
- Commerce - enough usage to produce a return on investment

Trump's Wall provides all of these.

But most importantly, President Trump's Wall weaves these technologies into an integrated whole, where they augment and support each other, creating a capability unique in human history.

Planned obsolescence. Just as the Erie Canal created conditions for its competition to be built, so too would Trump's Wall. And that is okay. We expect this. We plan for it.

Once we have a manufacturing presence in Earth orbit people would try to make other ways to get there. The Space Elevator mentioned before would be very cool.

Not to worry, Trump's Wall would not descend, wholesale, into obsolescence in an instant. It would be the hub of a transportation system for our nation and for the world for some time. It would be a huge, integrated, reconfigurable, manufacturing center. It would be a primary access to Earth orbit even with Space Elevators because it would be part of a much greater system. The space launcher may go obsolete soonest but all of the capabilities of the entire integrated system would provide too much value and be too difficult to easily replace.

It seems reasonable that one of the first things to do in orbit would be too generate energy in space. Earth orbit would allow for huge solar panel power generating arrays. The great advantage of orbital over terrestrial solar panels is there is no night, there is no atmosphere to attenuate sunlight and the panels can be always directed at the Sun.

We may also harvest energy directly from the Van Allen Belts by some clever means.

There has been much concern over how much Trump's Wall would cost and how to pay for it. And a simple "wall and a road" would cost a pittance compared to this proposal.

But actually a simple "wall and a road" would be much more expensive: financially, because it would not make any money, and politically because it would require funding, align with few personal interests and inspire no imagination.

When the human imagination has been captivated there is always enough money and manpower. There are always people who choose to risk and people who willingly give their blood, sweat and tears.

To say this project is too big, that we don't have enough money just isn't true. Money is always attracted to sound investments. Petronas planned to build a $36 billion Liquefied Natural Gas (LNG) project in British Columbia, Canada, until it was recently canceled for other than financial reasons. Tesla's 15 million square foot battery factory is expected to cost about $5 billion when completed in the next few years. There is always enough money.

There is always money for good investments. The question shifts from "How much does it cost?" to "Will the project pay for itself and return profits to its stock holders?"

Working and building in orbit will create a huge number of problems. People will apply themselves to the ones worth solving, inevitably producing a torrent of innovation and knowledge we cannot possibly anticipate. And where there are problems to be solved there is money to be made.

How do we handle space garbage? Clearing space of debris which can cause tremendous damage should it impact orbit facilities.

How do we handle space traffic control? The mathematics behind calculating orbits is quite complicated and very computing power intensive. The ability to anticipate orbital collisions will be critical.

And the list goes on and on, with every new development bringing a new set of challenges to be solved.

But, if we can build regular, reliable, economic access to Earth orbit it will quite possibly be the greatest wealth generator in human history. It will expand our horizons and open up possibilities undreamed of. It will change the way we see ourselves.

Conclusion

Please notice, nowhere in this submission is controlling our border mentioned. That is a side benefit of this economic engine.

Over the past decades there have been many good ideas that have all lacked one critical thing ... where to put them. Trump's Wall fills this need for all of them. More than that, located together they mutually support each other so the whole is greater than the sum of its parts.

- Transcontinental Superconducting Power Lines - We need these.
- Solar Panel Power Generation - With power lines to get the power where it is needed.
- Desalination Plants - Reliable/renewable fresh water supply.
- Transcontinental Railroad - A Panama Canal alternative.
- HyperLoop - Not just a fun thing, President Trump's Wall really needs it to be effective.
- Nuclear Power - Safe, secure, scalable, renewable.
- Space Launch System - Economically able to move enough mass into orbit to matter.

At present the public discourse is "How much will Trump's Wall cost?" Build this "wall" as an economic engine and the world will be pounding at our door demanding "How much can I invest?"

Incidentally, job creation in Mexico, due to solar panel power generation and fresh water land irrigation, would encourage Mexicans to stay in, and return to, Mexico.

Now, it is very easy to throw ideas around but actually implementing them can be daunting, to say the least. The number of private landowners and special interest groups for a project of this size must be mind-boggling.

But this would not be a static wall. This would be a much greater project designed to create many opportunities and vast wealth. This makes possible many potential solutions. As emotionally traumatic as having ones land built upon can be, it can also be a potential benefit. Just as over the past century oil drilling and modern fracking have brought some land owning families generations of income, so too could Trump's Wall bring generations of income to border land owners through leases and other agreements.

It has been said that no plan ever survives implementation. No one knows what the future holds. This book is about what may be, probably won't be but could be much better, if we only decide to try.

So how is that for a vision? Is it a worthy vision for a free people to pursue? Are we worthy to pursue it?

"Delays are the refuge of the weak minded, and to procrastinate on this occasion is to show a culpable inattention to the bounties of nature; a total insensibility to the blessings of Providence, and an inexcusable neglect of the interests of society...The overflowing blessings from this great fountain of public good and national abundance, will be as extensive as our country and as durable as time ... it remains for a free state to create a new era in history, and to erect a work more stupendous, more magnificent, and more beneficial than has hitherto been achieved by the human race."

- *De Witt Clinton*

(Closing paragraphs of the 1815 "Clinton's Memorial" addressed to the legislature of New York State petitioning for construction of the Erie Canal)

Further Thoughts

Commentary - Is Martian Colonization Realistic?

Perhaps one day we will colonize Mars. Perhaps one day we will even have the ability to go so far as to terraform Mars to meet our needs. But in order to do this we will need to control vast amounts of energy. The necessity of growing food pales in comparison to the challenge of creating an Earth-like environment. We must be realistic about our present capabilities.

Perhaps we could adapt to the 1/3rd gravity of Mars. The question is "how many generations of people are you willing to sacrifice to get that adaptation"? A thousand generations? More? Let us not forget that over half the population of Europe was wiped-out by the Black Death and still people have not acquired an immunity to it.

Mars has:

- 37.6% of Earth's gravity
    - 50% of Earth's diameter
    - 15% of Earth's volume
    - 15% of Earth's mass
- 0.67% of Earth's atmospheric pressure
    - An atmosphere of 95% carbon dioxide
    - An atmosphere 1% as deep as Earth's atmosphere
- 60% of Earth's solar irradiance
- No free water
- An average temperature of -63° C

- Zero protection from cosmic rays and high energy particles from the Sun.

Mars has a very weak and incomplete magnetic field and thus no meaningful protection from cosmic rays and the Sun's high-energy particles. Earth has a strong and complete magnetic field. The resultant Van Allen Belts deflect the harmful radiation, without which Earth would be burned to a lifeless rock. The Aurora Borealis are a visible representation of the high-energy particles from the Sun diverted by the Earth's magnetic field, beautiful but absolutely deadly should they ever reach the Earth.

The implications of the Martian environment to human life are dire. Human life has evolved in Earth's very specific environment. We are largely unaware, in our daily lives, of the aspects of our environment that are vital to our survival.

Human beings are creatures of a very limited, specific environment. Many of the conditions necessary for life are so much a part of our environment we are not even aware of them in our daily lives. We cannot feel them. We rarely feel atmospheric pressure. We cannot feel the atmosphere we breathe. We do not think of the gravity that is ever present. And yet without them we cannot survive.

We are blissfully unaware of the consequences a non-Earth-like environment will have on the ongoing cascade of chemical reactions within the human body. However, there has been some research done.

> *"So what does this all mean for a mission to Mars and for NASA's future plans for deep space exploration? Our recent and current data, along with data from other laboratories, have now clearly identified an unexpected sensitivity of the CNS [Central Nervous System] to cosmic radiation exposure. With this realization comes a concomitant understanding of the risk for developing cognitive deficits that may predispose astronauts to performance decrements, faulty decision-making and longer-term neurodegenerative effects. Given the increased demands, uncertainties and variety of stressors inherent in deep space travel, defining acceptable risk specific to radiation exposure remains a challenge."*
> *- P. K. Vipan, et all*
> https://www.nature.com/articles/srep34774

No doubt we can develop living environments with adequate radiation shielding. But that is not all. As we see in saturation diving compression chambers, various microbes change in character under different pressures and gasses creating things that humans have not evolved immune defenses against.

Unavoidably, there will be unanticipated and unintended consequences of artificial human habitats under various conditions like low gravity.

Like most of our solar system, the Martian environment is completely antithetical to human life. Successful human colonization of Mars will require the creation of a close to Earth-like environment, and this will require the control of immense amounts of energy, far beyond present human capability.

At present, attempting human colonization of Mars seems to be a fools errand, the costs far outweighing any conceivable benefits.

Commentary - Martian Colonization Governance Is It Realistic?

If the purpose of Martian colonization is to "save mankind" when it destroys itself on Earth, let us stop to consider.

One of the primary mechanisms of maintaining power used by totalitarian regimes is the forcible detainment of its citizens. Not only are the people not allowed to leave their country but their travel and residence are determined by the government. And this is on Earth, where the land is radiation free and where there is air to breathe, even if restricted by walls and wire.

However in the completely hostile environment of Mars, where the only habitable spaces were fully enclosed and sealed man-made structures, where "outside" is not just too cold or too hot or too wet or too dry but is immediate death, people's ability to escape tyranny would be much more restricted.

Mars is awash in burning radiation, poisonous gasses, cell destroying near vacuum and destructive temperature differences. It is a cruel place where determination, fortitude and toughness will not sustain justice and the right when there is no air to breathe.

We must be careful not to confuse righteous anger for ability.

What would such a society look like? Would it be populated by independent free men? Would that be possible? Would not the reality of central control of the very air people breathe be too tempting? Would this not create conditions that select for those most capable of acquiring power and most brutal in its application? The natural result of which being a dystopian Orwellian nightmare?

Perhaps there is a way. Maybe. Perhaps if each household had its own independent means of power generation? It is not clear to me.

But one thing is certain. Those who go had better think about this and prepare carefully beforehand.

One of the primary reasons Jim Jones moved his cult to an inaccessible jungle in a third-world country was because it made it impossible for the members of his cult to leave. The people went willingly, with high hopes, but the realities of the environment created the conditions that made individual rights impossible and selected for tyranny. The tragedy of Jonestown was the result.

An environment where people have the ability to leave at any time selects for cooperation … because otherwise people can just take their toys and leave.

Notice the distinction between "the ability to leave" and "the option to leave". "Option" implies something given or bestowed by another, in this case the government. And what government giveth it will taketh away … eventually.

Any human colonization located anywhere hostile enough to require people to live in fully enclosed and sealed man-made structures had better have the technology and understanding to intentionally create a physical environment that promotes individual liberty or the inhabitants will pay a terrible price in human suffering.

Commentary -Unabashed American Exceptionalism Do we still have it?

Here is the title of an article that caught my eye.

"'Elon Musk's Hyperloop might not debut in the U.S. — and that's a good thing'
So says outgoing U.S. Secretary of Transportation Anthony Foxx"... (January 16, 2017)

> *"The technology, the science behind it, is very sound,"*
> *he said. "But it's one of those examples of, the*
> *technology may be there before the government is."*
>
> *"Foxx said applying current railway regulations to*
> *Hyperloop 'would be like putting a square peg in a*
> *round hole.'"*
>
> *"Hyperloop would require Congressional action before*
> *the DoT [Department of Transportation] can "jump*
> *into it with two feet" and start making rules, he noted."*
>
> https://www.recode.net/2017/1/16/14280836/elon-musk-
> hyperloop-anthony-foxx-high-speed-rail-train-transportation-
> recode-podcast

These comments from past Secretary of Transportation Anthony Foxx are embarrassing, or should be to all Americans. This attitude represents the strangulation of the

American "can do" spirit by bureaucratic malaise. It is the American people's confidence and belief in their ability to create a better future that made the United States of America the most successful, most free, wealthiest society in the history of mankind. The day the majority of the American people adopt this helpless attitude and allow government to stifle creative action then history will mark that as the date the U.S. entered its decline.

Foxx displays the attitude of a comfortable, timid, small-minded bureaucrat. This has no place in the building of America. It does not speak of unabashed American exceptionalism. It is the antithesis of what built the USA. Does history show that the men who built the United States of America sat passively by saying "That is a fine idea but let's wait until someone else does it first.", "Let's wait until the government 'makes the rules' and 'gives us permission.'"? No they did not. Not Vanderbilt, Not Carnage, Astor, LeTourneau, Ford, Westinghouse, Tesla, Edison, Burr, Roebling, Hill nor Kaiser. Whatever opinion one holds of these men, one thing is for certain, they did not wait around for someone to give them permission. They saw opportunities and acted, and America and the world are wealthier and better for it.

"Civilizations die from suicide, not from murder"
- Arnold Toynbee

Commentary - How Developed is Mankind?

The Kardashev Scale
        https://en.wikipedia.org/wiki/Kardashev_scale

- **Type I civilization** - also called a *planetary civilization* - can use and store all of the energy which reaches its planet from its parent star.
- **Type II civilization** - also called a *stellar civilization* - can harness the total energy of its planet's parent star (the most popular hypothetical concept being the Dyson sphere - a device which would encompass an entire star and transfer its energy to the planet(s)).
- **Type III civilization** - also called a *galactic civilization* - can control energy on the scale of its entire host galaxy.

This scale is hypothetical, but notice it is based on energy consumption. More to the point, useful energy consumption. Energy put to useful work meeting the needs of the civilization.

Perhaps it has not escaped you that, on the Kardashev Scale, mankind **doesn't even register!**

Mankind is pretty insignificant compared to the titanic forces constantly swirling around us, most of which we are completely unaware of as we live our daily lives.

I find it a little depressing, somewhat frustrating but mostly infuriating. One would hope that it would instill a dash of humility but mostly I am just impatient for progress.

The relative magnitude of energies involved, just here on Earth, piqued my curiosity. So I ran a few basic calculations.

Hawaii's Kilauea Volcano has been erupting continuously since 1983. Around 2016 I saw a video of bright-orange lava pouring into the ocean from Kilauea.

The video showed a lava "waterfall" pouring from a cliff into the ocean. From the video I roughly scaled it to obtain dimensions in order to determine how much energy it contained.

66 feet: Height of "waterfall"
10 feet: Diameter of lava stream
10 feet: Horizontal distance from the cliff to hitting the water.
Orange: Color of lava stream (900 deg C)

From this follows:
Time of fall for lava to hit the water:
$d = 1/2 at^2$ therefore $t^2 = 2d/a$
$t(seconds) = sqrt\{(2)(66\ ft)/(32.2\ ft/sec^2)\}$ = ~2 seconds

From this we find the lava flow rate:
10 feet of horizontal travel in 2 seconds
$d = vt$ therefore $v = d/t$
Velocity (feet/second) = 10 ft/2 seconds = 5 feet/second

Volume of lava per time:
Assume a circular lave stream with a 10 foot diameter
Cross-sectional Area = $Pi(r^2) = (3.14)(5^2)$ =~ 80 square feet
80 square feet x 5 feet/second =

400 cubic feet of lava per second or
11.3 cubic meters per second

Basalt weighs about 3,000 kilograms per cubic meter
(3,000)(11.3) = 34,000 kilograms per second or
The lava flow is about 17 tons per second.

Hawaiian lava is Basalt.
Liquid Basalt shows its temperature by its color.
Bright Red: 700 deg C
Orange: 900 deg C
White-Hot: 1,200 deg C

Basalt specific heat is 0.84 kJ/kgK
K is temperature change for which we will use 900
(0.84)(34,000)(900) = 25,704,000 kiloJoules per second (kJ/s)
kJ/s = kiloWatt (kW)

Therefore this one lava flow embodies:
26,000,000 kW x 24 hours per day = 624,000,000 kWh/day
228,000,000,000 kWh per year or
228,000 GigaWatt hours per year (GWh/year)

World energy consumption is about
176,000,000 GWh/year

Therefore:
228,000/176,000,000 = 0.13%
This one small lava flow embodies 0.13% of total human energy consumption.
About 772 of these lava flows would equal total human energy consumption.

U.S. energy consumption is about 30,000,000 GWh/year
228,000/30,000,000 = 0.76%
This one small lava flow embodies 0.76% of total U.S. energy consumption.
About 132 of these lava flows would equal total U.S. energy consumption.

Energy in a barrel of oil is 1,700 kWh
U.S. oil production is about 12,000,000 barrels per day or
4,380,000,000 barrels per year
(1,700)(4,380,000,000) = 7,446,000,000,000 kWh/year or
7,446,000 GWh/year

228,000/7,446,000 = 3%
About 33 of these lava flows would equal the energy in the total U.S. oil production.

This one small lava flow embodies:

- 0.13% of total human energy consumption.
  - About 772 of these lava flows would equal total human energy consumption.
- 0.76% of total U.S. energy consumption.
  - About 132 of these lava flows would equal total U.S. energy consumption.
- 3% of total U.S. oil production.
  - About 33 of these lava flows would equal the energy in the total U.S. oil production.

To say that "Kilauea Volcano has been erupting continuously since 1983" is misleading. No doubt this is an accurate statement concerning this one particular lava outpouring we have come to know as Kilauea Volcano, but perhaps perspective could be gained from a larger view.

Hawaii is a chain of volcanic mountains. The tallest of which, where as it is 13,800 feet above sea level, is in excess of a staggering 33,000 feet above the sea floor. Mount Everest is 29,029 feet above sea level. And Mauna Kea is only one in this chain of volcanic mountains.

To create Hawaii's 1,500 mile long mountain chain, vast quantities of lava have been pouring out of the earth for the last 28 million years or so.

Alaska's Aleutian island chain has 57 volcanoes, Indonesia has 400 volcanoes, and the list goes on.

The 2011 giant tsunami that wrecked havoc in Japan and elsewhere was the result of an underwater earthquake. The earthquake was a result of the movement of two tectonic plates, one sliding over the other.

The amount of energy released during this earthquake has been estimated at 10,000,000,000,000,000,000,000,000 kWh.

This is about 62.5 billion times more energy than mankind consumes in an entire year, and the earthquake released it in less than 15 minutes.

Mankind's 2016 world energy consumption was about 176,000,000,000,000 kWh.

The amount of solar energy that reaches the Earth's surface per year is (assuming about 30% is reflected back into space) about 1,100,000,000,000,000,000 kWh.
          https://en.wikipedia.org/wiki/Solar_energy

Therefore, the Earth receives 6,250 times more energy from the Sun than we command.
We control 0.016% of the amount of energy the Earth receives from the Sun.
The Earth receives more energy from the Sun in an hour and a half then mankind uses in a year.

Pathetic! Mankind is a long way from being even a Type I (planetary) civilization.
Let's get cracking!

The sun produces about
3,380,000,000,000,000,000,000,000,000 kWh per year.

The disk of the Earth intersects about 2 one-billionths of the sphere around the Sun at its orbital distance. About one one-billionth of the Sun's total energy output reaches the Earth's surface.

The Sun produces about 19 trillion times more energy than mankind controls. Mankind controls about 0.000000000005% of the amount of energy output of our Sun.

We are so insignificant to a Type II (stellar) civilization it is laughable.

Commentary - How Far Has Man Come?
How much further will we go?

We humans have come a long way, from bare huddled survival in a merciless world to a population of billions. Undisputed masters of the animal kingdom, control of mechanical power and even access to space. To have come from where we began, human accomplishments are almost unbelievably impressive ... and yet we know so little.

Our control of nuclear power is an almost unimaginable feat for a naked ape. It is a testament to human curiosity and ingenuity, but we use it to boil water. Oh, we put it to useful work by using the steam to turn steam turbines to generate electrical power, but, not withstanding impressive advancements in turbine design, materials science, and manufacturing, to us, a nuclear reactor is essentially a "high-tech" campfire.

In all of human knowledge, of the forces of nature we know of only four.
1. Electromagnetism
2. The Strong Nuclear Force
3. The Weak Nuclear Force
4. Gravity

Looking at this list we must face the fact that we have very little control of nature. We are in the infancy of our understanding.

True, we have developed an almost god-like mastery of electromagnetism. We routinely and accurately control from the huge amounts of power in continent spanning electrical grids to the tiniest trickles of electricity in computer chips and electron microscopes. With seeming ease we move energy between different forms; heat, electrical, mechanical, etc. We power our industry and fertilize our fields by the energy we store in the chemical bonds of the fuels and fertilizers we make on an industrial scale. With delicate precision we condition electricity into different forms, frequencies and magnitudes. We even have the ability to manipulate single electrons and create collimated light.

By putting the electromagnetic force to useful work for us billions of people have been elevated from the back-breaking, soul-crushing physical drudgery of pre-industrial life. And this is an amazing accomplishment.

However, as to the Strong Nuclear Force and the Weak Nuclear Force we know precious little and have even less ability to practicably apply them in our service.

And as to the fourth, Gravity, our lack of understanding is so complete we don't even know what questions to ask.

Yes, we can fly, but only by counteracting the force of gravity, not by manipulating it. The interaction of air molecules, that cause lift when air is forced to flow over a wing, are a result of electromagnetic forces. As is chemistry, the chemistry of combustion to produce thrust.

We defy gravity, we don't control it. We have no theory of quantum gravity. We lack fundamental understanding. And without that, we have no pathway toward specific knowledge and control.

My, we have a long way to go.
But that makes life interesting … if we want it to be.

Commentary - Committees Kill Genius

I am not interested in design by committee. That is the surest way to produce the bland and uninspired. No, the work must be the inspiration of one driven genius mind. But it must not be given carte blanche. It must be overseen, there must be final arbiters who require it to justify every concept and fight for every detail. Designs must be created and destroyed and new ideas must rise from the ashes by shear force of will, a will that refuses to surrender. The final design must be forged in the fires of adversity, competition and struggle. The process must not be made easy. It must not be made comfortable. Adversity and competition is the only way to create a result that will make us proud.

New York's Grand Central Station is a magnificent structure. It is a near perfect balance of space and structure, an embodiment of culture, grace and beauty down to the finest details. Not only that, it effortlessly blends revolutionary functionality with inspiring, uplifting, grandeur. It wraps the most modern technological advances of its day in an unabashed celebration of the beautiful proportions of several thousand years of our cultural heritage. The whole effect uplifts the human soul, imbuing travelers with a sense of purpose and possibilities. It is not only the finest train station in existence, it remains the greatest station of any type in the world.

Why? How is it that it came to be? On the surface, the story of the building of Grand Central Station is uneventful to the point of boring. No one died in its construction, there were no delays, there were no scandals, it was built to plan, reasonably on time and within budget.

But its magnificence and brilliance was forged in the fires of a competitive environment.

This is a tale of two firms. Two forces of genius that each alone would have produced the unremarkable, but forced to find common ground they produced one of the true wonders ever created by mankind.

How was this accomplished?

Design rarely leaps fully formed into final form. It is most often a painful, frustrating, effortful process of creation and destruction and creation again. So too, the product of one mind often benefits from externally imposed discipline. Grand Central Station is a classic example of this.

The board of directors of the New York Central Railroad focused two strains of geniuses upon their goal. They gave them great creative freedom but mercilessly held them to account, acting as final arbiter.

The first, the architectural firm of Reed and Stem, were brilliant innovators in the area of traffic patterns and efficiency in moving people. Focused on the vital practical aspects of its function, their first attempt was a bland functional structure over an ingenious transportation system.

The board of directors was less than impressed. They then contributed their genius by creating a competitive environment that forced discipline into the process by introducing a countervailing force of equal strength and vigor.

This second force was Whitney Warren, a classically trained architect of uncommon ability.

Both parties were brilliant, opinionated rascals who loathed interference in their work. But the opportunity to put their personal stamp on such a prominent structure kept them engaged, arguing over every last detail.

Deeply contentious, but what a result! From the battle rose uncommon magnificence, far superior to what either could have produced alone.

Whitney Warren's magnificent vision defines the building, it is true, but without Reed and Stem's innovative designs in effortless flow patterns for moving people it would not have functioned smoothly as a whole transportation system. Resistance and frustration would have been introduced preventing people from being inspired by the detail, proportion and spaces.

Why did the Soviets or Communist Chinese never build anything that rivals New York's Grand Central Station. Because they designed by committee. Their process was designed specifically so no one made any decision, so no one was responsible and therefore no one cared. Totalitarian States build to dominate, to glorify their power, to frighten their people. Their results are often impressive but always oppressive. They make people feel small and weak. Built, not for the people and by the people but, to subjugate the people, they crush the spirit rather than uplift the soul.

Let us create an opportunity for the rare genius to create for us, it will make us proud and we shall all be better for it.

What does this mean? We need to consciously create competitive environments that select for genius and allow it to flourish.

Create prizes.

Some of the most remarkable human achievements have been the results of prizes. These competitions provide reward in the form of wealth and celebrity, providing incentive to risk and sacrifice and strive to those who choose to dare. And interest and excitement for those who choose to watch.

Charles Lindbergh was catapulted to instant world fame by winning the Orteig Prize for his daring trans-Atlantic Ocean solo airplane flight.

John Harrison struggled for over 20 years to win the prize offered by the British Parliament for a practicable solution to one of the major problems of his day, the problem of longitude. Thus society benefited from his great advances in accurate timekeeping.
        https://en.wikipedia.org/wiki/John_Harrison

And the list goes on and on.

With respect to President Trump's Wall, consider that it need not be identical its entire length. Yes, certain aspects need to be consistent for pipelines, power lines, rail lines, etc and standards are not only useful they are vital to making an integrated whole. But there is still much creative latitude possible. Encouraging different ideas, through prizes, would create the competitive conditions for the best solutions to rise to the top.

Genius is a rare and precious resource. We must support it, demand discipline and inspire greatness from it by providing important challenges and lavishing great reward upon success. To do anything less belittles us all.

Commentary - If Not the USA, Who?

What nation, to you, would be acceptable to hold power over all humanity?

If the United States of America can not muster the resolve and fortitude to build Trump's Wall and thus dominate Earth orbit, who will fill the vacuum?

Ask yourself, "If not the U.S., what country could build it?" Try it. Imagine [*insert your choice here*] building it. Is the idea immediately ridiculous? For most countries it is.

Pick a country you think could build Trump's Wall and economical access to space, say Russia. Try to imagine what the end result would be. A centrally controlled, dystopian, Orwellian, totalitarian nightmare. A massive gulag used to subjugate its own people and threaten the world.

Not the Russians? Perhaps the Chinese? Pick a nation both capable and worthy, convince me. The United States is imperfect to be sure but if not us, who?

No, the only people who could build Trump's Wall and use it to improve people's lives and move human understanding forward are a free people engaged in free enterprise. All other societies would quickly find it an instrument of oppression used on them rather than for them.

Make no mistake, the world is now entering a space race. Not by choice and most people are blissfully unaware, but it is reality.

And this is not the space race of our grandfathers, where bragging rights were largely what was at stake. No, today the stakes are infinitely higher.

Humanity is at a crossroads. The brutal reality is whoever gets to space first with the most and dominates Earth orbit will irrevocably determine the course of humanity. And in-spite of America's imperfections, wouldn't all mankind be better off lead by a free people?

One thing is for certain ... if the United States of America doesn't build the capability of Trump's Wall and use it vigorously, some other people will, and what would the world look like then?

Commentary - The Real Prize
is economical access to Earth orbit

Regular, reliable, economical access to Earth orbit is the real prize, as it will enable manufacturing in space which will advance human understanding and create great wealth.

Working and building in orbit would create a huge number of problems. People would apply themselves to the ones worth solving, inevitably producing a torrent of innovation and knowledge. We have no way of predicting what knowledge, advancements and innovations would come from many people competing in a vibrant low gravity, near vacuum manufacturing environment, but come they assuredly would. And where there are problems to be solved, there is money to be made.

Furthermore, I believe regular, reliable, economical access to Earth orbit would be the greatest wealth creator in human history. It would create millionaires like hot cakes. And all Americans would be able to participate. Yes, some few as business owners, inventors, etc., would make vast fortunes but, you say, that would be a fairly small percentage of people, wouldn't the rest of us be left behind?

Well, consider this, what would your finances look like today if you had purchased MicroSoft stock in the 1980's?

Yes, some would get richer than most but the vast wealth creation would create a huge number of opportunities where even the least able would have choices. The difference in results would, as always, be in the choices and actions each individual took, or did not take.

How many millionaires did Bill Gates create on his way to becoming a billionaire? And, yes, some of those MicroSoft millionaires were secretaries, secretaries who made good choices.

Does the fact that Bill has more money than I do make my life worse? I don't think so.

Did he gain his billions by taking from others? No. Regardless of the opinion one may hold of him, he created wealth through free exchanges with other people and in so doing made their lives better. People's lives improved, he grew rich and the total economy grew.

I reject the idea that the world is "zero sum". That there are a finite amount of resources and therefore in order for one to get rich he must take from another. I choose to believe human ingenuity makes the world infinitely bountiful.

The question is:

Are we going to see other people's good fortune as a grievance, or use it as an incentive for self-improvement?

I, for one, choose the latter. I hope you choose to come along with me.

Commentary - A Man's Reach Should Exceed His Grasp

Wow. When we stop for a minute to consider, President Trump's Wall may seem impossibly huge. The very prospect may seem terrifying.

I remember walking into a professional sports stadium for the first time. Standing on the ground floor looking up at the ceiling, the scale being so huge, the feeling of "fear of falling" washed over me ... so strange.

And so it is, from time to time, for me with this project. Sometimes the magnitude of it rises up before me. But that effect is transient. What is left unaffected, undiminished, is my absolute belief that not only can this be done, it is mankind's destiny to build it. To gain regular, reliable, economical access to Earth orbit. It is just waiting for a people, with the myriad required specialized skills and knowledge certainly but mostly, with the will. The will to solve problems, persevere through hardship, to strive today for a greater tomorrow.

For this is what mankind is destined to do, to push our boundaries, to expand our capabilities, to seek out the unknown.

> *Ah, but a man's reach should exceed his grasp, Or what's a heaven for?*
> *- Robert Browning*

Afterwards

Government has its right and proper place in society but it is worse then bad at building. Government's fundamental structure encourages wastefulness, low quality and corruption in construction projects. It has an equally poor track record in managing real estate and running businesses.

Furthermore, It is not in our national interest to create something that will become a political weapon, a bone of contention in every federal budget cycle.

The key to avoiding these issues is not to try to legislate a solution but to keep the government out of it as much as possible.

Therefore, build President Trump's Wall with private funds to the greatest extent possible. It has been said that better decisions are made by folks exercising common sense in the spending of their own money than any government committee composed of the "best minds".

> *"When it comes to making a smart decision, the most distinguished planning committee working with the highest-priced consultants doesn't hold a candle to a group of guys with a reasonable amount of common sense and their own money on the line."*
> *- Donald Trump*
> *Trump: The Art of the Deal p. 283*

To say that we must have greater scrutiny to see that our tax dollars are spent wisely is fighting the wrong battle.
No one spends someone else's money as carefully as he spends his own. And the worst case is when he is spending someone else's money for the benefit of others.

This fight is not winnable. Spending oversight is an ineffectual, clumsy way of trying to override human self-interest.

Creating a system that puts desired results in conflict with human nature is foolish.

The only sound solution, one that allows human nature to actively pursue the desired results, is when someone is spending his own money in his own self-interest.

Individuals with their own money on the line have a strong self-interest in controlling costs, meeting deadlines and providing a product or service that people want at a price people are willing to pay.

The bottom line is private enterprise projects are always, and in all ways, superior to those attempted by government.

In order to do this the project must be attractive to private investment. It must make money early and often in its development. The role of government in this endeavor is to create an environment and then get out of the way, save as referee.

But grand plans are nothing without human will, initiative and drive, the human spirit of rising to a challenge. Does our culture still have people that desire? Are we still a people with confidence, ability and purpose? Do we have those among us who are driven to dare, to risk - for the reward certainly - but mostly for the doing of it, for the battle, for the striving, for the living and because they can't not? Everywhere I have worked across this great land there have been at least some. I, for one, choose to believe in the indomitable American spirit.

> *"But most of all, I was remembering how my answer to my own doubts, every time, was my faith in my country. I have always believed that America could accomplish anything she set out to do."*
> *- Jon van Hardeveld*
> *(Panama Canal engineer during construction)*

People can be judged by the challenges they accept. How will we, as a people, be judged?

If you say you can't build it ... I will believe you.
But kindly do not presume to speak for the rest of us.

www.ingramcontent.com/pod-product-compliance
Lightning Source LLC
Chambersburg PA
CBHW070930210326
41520CB00021B/6870